1st International Conference on Ceramic Glass and Soft Matter (ICCGS 2020)

IOP Conference Series: Materials Science and Engineering
Volume 1160

Fez, Morocco
11 - 12 November 2020

Editors:

Salaheddine Sayouri
Yahya Ababou
Mohammed Filali
Abdelilah Rjeb
Mohamed Naji

ISBN: 978-1-7138-3427-4
ISSN: 1757-8981

Printed from e-media with permission by:

Curran Associates, Inc.
57 Morehouse Lane
Red Hook, NY 12571

Some format issues inherent in the e-media version may also appear in this print version.

Institute of Physics
Dirac House, Temple Back
Bristol BS1 6BE UK

Phone: 44 1 17 929 7481
Fax: 44 1 17 920 0979

techtracking@iop.org

Additional copies of this publication are available from:

Curran Associates, Inc.
57 Morehouse Lane
Red Hook, NY 12571 USA
Phone: 845-758-0400
Fax: 845-758-2633
Email: curran@proceedings.com
Web: www.proceedings.com

TABLE OF CONTENTS

STRUCTURAL AND DIELECTRIC PROPERTIES OF PB (ZR $_{0.52}$ TI $_{0.48}$)$_{1-3X/4}$ FE $_{X}$ O_3 CERAMICS AT 0≤X≤0.20 PREPARED BY SOL-GEL METHOD 1
Ahabboud Malika, M Amarass, F Z Ahjyaje, F Abdi, T Lamcharfi

PROBING THE BALANCE OF TELECHELIC POLYMERS BRIDGING AND SCREENED-COULOMB INTERACTIONS IN MICROEMULSION SYSTEM 10
R Ahfir, S Elkhaoui, M Khatouri, M Lemaalem, L Talha, M Filali

AN APPROACH TO DETERMINE THE DIATOMIC VIBRATIONAL ENERGY LEVELS: APPLICATION TO LIH, NO AND CO MOLECULES 17
I Amila, M Janati Idrissi, A Fedoul, S Sayouri

LOW TEMPERATURE TREATMENT AND STRUCTURAL CHARACTERIZATION OF NA $_2M_2FE(PO_4)_3$ (M= MN OR NI) ALLUAUDITE PHASES 23
Nour El Hoda Bouftila, Mohamed Naji, Yahya Ababou, Abdelilah Rjeb, Salaheddine Sayouri

SOL-GEL SYNTHESIS AND STRUCTURAL STUDY OF A LITHIUM TITANATE PHASE LI $_{3X}$LA $_{2/3-X}$ [] $_{1/3-2X}$ TIO_3 AS SOLID ELECTROLYTE 32
Abdelhak Chouiekh, Mohamed Naji, Abdelilah Rjeb, Yahya Ababou, Salaheddine Sayouri

DIFFUSE AND RELAXOR PHASE TRANSITIONS OF BA $_{0.95}$ BI $_{0.05}$ TI $_{1-X}$ FE $_{X}O_3$ CERAMICS AT X=0.00 TO 1.00 OF FE CONTENT, PREPARED BY SOLID STATE METHOD 41
Gouitaa Najwa, Lamcharfi Taj-Dine, Abdi Farid, Echatoui Nor-Said, Amarass Mustapha

EFFECT OF DIELECTRIC SCREENING FUNCTION ON BINDING ENERGY AND DIAMAGNETIC SUSCEPTIBILITY OF IMPURITY DOPED IN GAAS CONICAL QUANTUM DOTS 54
E Iqraoun, A Sali, K El— Bakkari

OPTICAL ABSORPTION STUDY OF MANGANESE-DOPED NA $_2$O-K $_2$O-P $_2$O $_5$ GLASSES 63
M Jerroudi, L Bih, M Haddad, I Saadoune

STRUCTURAL PROPERTIES OF OIL/WATER MICROEMULSION WITH SURFACE GRAFTED PEO POLYMERS 69
M Khatouri, R Ahfir, S Elkhaoui, M Lemaalem, L Talha, M Filali

FIRST PRINCIPAL CALCULATION AND MONTE CARLO SIMULATIONS OF THE MAGNETOCALORIC EFFECT, ELECTRONIC AND MAGNETIC PROPERTIES IN PEROVSKITE OXIDE PR $_{0.65}$ SR $_{0.35}$ MNO $_3$ 77
G Kadim, R Masrour, A Jabar, E K Hlil

STRUCTURAL AND OPTICAL PROPERTIES OF CDO-ZNO THIN FILMS SYNTHESIZED BY SOL-GEL SPIN COATING METHOD 84
A Nfissi, M Belhajji, L Mrharrab, Y Ababou, S Sayouri

Author Index

IOP Conf. Series: Materials Science and Engineering 1160 (2021) 011001 doi:10.1088/1757-899X/1160/1/011001

Preface

The 1st International Conference on Ceramic, Glass and Soft Matter (ICCGS 2020) is an event that brings together specialists from all over the world, covering multiple aspects of diverse topics such as: Dielectrics, Multiferroics, Glass, Polymers and Colloids and Numerical Simulations.

Due to the COVID-19 outbreak, the ICCGS 2020 conference, which was supposed to be held in March 2019 in Fez, Morocco, was organized on-line using Google Meet Platform on 11-12th November 2020.

The organizing committees, session chairs have made a lots of efforts to make this 1st edition successful. All of them were present in Fez and were coordinating during these two days of the conference. Thank you very much for your support and contributions. We could not have pulled off this conference without all of your hard work and dedication.

The conference received submission from different countries, with the main track receiving 102 submissions. It was consisted of 4 Plenary sessions (from USA, France, South Africa and Morocco), 34 oral presentations (including presentations from France, Algeria and Morocco) and 2 Poster sessions.

The welcoming ceremony was live-streamed on Facebook, and have reached more than 1000 views (participation). This session has known the participation of several members of the honorary committee as well as representatives of the Crafts Chamber which have opened the debate on the importance of Ceramics, Glass and Soft-Matter to the development of the Economy of Fez-Meknes region as well as the urgent need for the creation of a 'pool' of skilled students, engineers and researchers in these areas.

Each plenary talk lasted for 45min including 5min for Q&A, while for the authors' presentation, 15min including Q&A was allocated for each. Two parallel sessions were held together with each having about 40 participants in the audience. All participants were active, questions raised after every talk. The whole conference was held online in Google Meet and for each session the audio-

IOP Conf. Series: Materials Science and Engineering 1160 (2021) 011001 doi:10.1088/1757-899X/1160/1/011001

video quality was checked with each speaker. For authors having some technical problems, we also had pre-recorded video presentations as an alternative for online presentation.

As a conference chairs, we know that the success of the conference depends ultimately on the many people who have worked with us in planning and organizing both the scientific and technical program. In particular, we thank the Program Chairs for their wise advice and brilliant suggestion on organizing the technical program; the conference reviewers for their thorough and timely reviewing of the papers, and our sponsors who have helped us to keep down the costs for all participants. Recognition should go to the Organizing Committee members who have all worked extremely hard for the details of important aspects of the conference programs and activities. Lastly, we would like to thank our invited speakers who have agreed to address the conference attendees and we thank all of the conference participants for their contributions, which are the foundation of this conference.

ICCGS 2020 Conference Chairs

IOP Conf. Series: Materials Science and Engineering 1160 (2021) 011001 doi:10.1088/1757-899X/1160/1/011001

Conference Scope

The International Conference on Ceramic, Soft Matter and Glass is an interdisciplinary forum on the most recent achievements on the physics and chemistry of Ceramics, Soft Matter and Glasses. The conference brought together leading academic scientists, researchers and scholars in the domain of interest from around the world. Topics of interest include: Dielectric, Ferroelectric and Magnetic materials, Polymers, Colloids, Microemulsion, Glass Structure and Chemistry, Glass Physics and properties, Computer simulation, Molecular Dynamics.

TOPICS

- Ceramics, crystals.
- Dielectric , Ferroelectric, Piezoelectric and Magnetic materials.
- Ferroelectric, relaxor and multi-functional ceramics.
- Polymers, Gels, Colloids.
- Chemistry and Physics of Glasses.
- Biological Physics, Macromolecular Crowdin
- Numerical simulation methods.
- Molecular Dynamics

IOP Conf. Series: Materials Science and Engineering 1160 (2021) 011001 doi:10.1088/1757-899X/1160/1/011001

Honorary Committee

- **Pr. Radouane MRABET**, President of USMBA Fez, Morocco
- **Pr. Brahim AKDIM**, Vice-President of USMBA, Fez Morocco
- **Pr. Abdelrhani ELACHQAR**, Vice-President of USMBA, Fez Morocco
- **Pr. Mohammed BENLMIH**, Dean of FSDM - USMBA, Fez Morocco
- **Pr. Mohammed EL HASSOUNI**, Vice-Dean of FSDM - USMBA, Fez Morocco
- **Pr. Mohammed BELKASMI**, Vice-Dean of FSDM - USMBA, Fez Morocco

Conference Editors

- Pr. Salaheddine Sayouri (FSDM – USMBA, Morocco)
- Pr. Yahya Ababou (FSDM – USMBA, Morocco
- Pr. Mohammed Filali (FSDM – USMBA, Morocco)
- Pr. Abdelilah Rjeb (FSDM – USMBA, Morocco)
- Pr. Mohamed Naji (FSDM – USMBA, Morocco)

IOP Conf. Series: Materials Science and Engineering 1160 (2021) 011001 doi:10.1088/1757-899X/1160/1/011001

Organizing Committee

Conference Chairs:

- Salaheddine Sayouri (FSDM – USMBA, Morocco)
  salaheddine.sayouri@usmba.ac.ma
- Mohammed Filali (FSDM – USMBA, Morocco)
  mohammed.filali@usmba.ac.ma
- Mohamed Naji (FSDM – USMBA, Morocco)
  mohamed.naji3@usmba.ac.ma

Technical Program Committee Chairs:

- Yahya Ababou (FSDM – USMBA, Morocco)
- Abdelilah Rjeb (FSDM – USMBA, Morocco)
- Salma Iraqi (FSDM – USMBA, Morocco)
- Abdelaziz Fedoul (FSDM – USMBA, Morocco)
- Boujemaa Jaber (CNRST, Morocco)
- Brahim Fakrach (FS Meknes– UMI, Morocco)
- Abdelhafid Azougarh (FSDM – USMBA, Morocco)
- Ismail Boumhidi (FSDM – USMBA, Morocco)
- Saad Bensouda (USMBA, Morocco)

Student Committee

Mohammed Khatouri
Nour El Houda Bouftila
Aziz Nfissi
Soumia Elkhaoui
Fouzia Saadaoui
Mounir Belhajji
Abdelhak Cheouikh
Adil Moutawaffiq

Web Master:

Mohamed Naji

IOP Conf. Series: Materials Science and Engineering 1160 (2021) 011001 doi:10.1088/1757-899X/1160/1/011001

Scientific Committee

IOP Conf. Series: Materials Science and Engineering 1160 (2021) 011001 doi:10.1088/1757-899X/1160/1/011001

PLENARY SPEAKERS

Mario Maglione, *Research Director at Institut de Chimie de la Matière Condensée ICMCB, UMR 5026 CNRS, University of Bordeaux.*

Research Interests: *Dielectrics, Piezoelectrics, Thin films, Interfaces.*

Plenary talk title:

Universal Behaviour of BaTiO3-BaZrO3-CaZrO3 ceramics, single crystals and thin films

---

Vijaya Srinivasu Vallabhapurapu, *Professor at College of Science, Engineering and Technology, University of South Africa.*

Research Interests: Experimental Condensed Matter Physics, Electron Spin Resonance (ESR), Nano Magnetics, Thermal Properties of Nano and Micro Composites, Spintronics.

Plenary talk title:

Low field microwave absorption and magneto resistance in Pani-Fe Nano fibre composites.

IOP Conf. Series: Materials Science and Engineering 1160 (2021) 011001 doi:10.1088/1757-899X/1160/1/011001

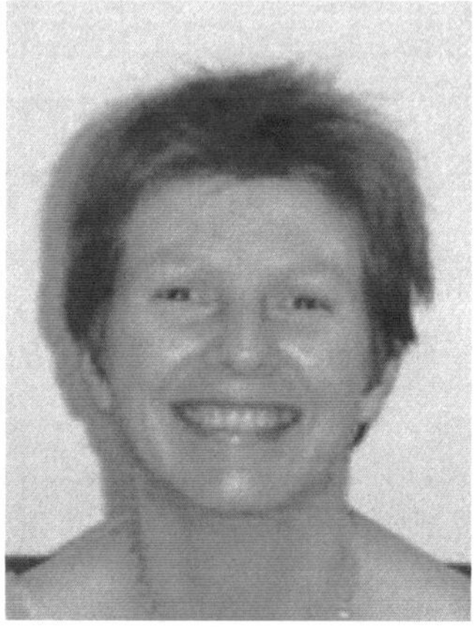

Laurence Ramos, *Research Director at Charles Coulomb Laboratory (L2C),UMR 5221 CNRS, University of Montpellier.*

Research Interests: Soft Matter, Jamming, Plasticity, Materials Failure and Rheology

Plenary talk title:

Role of normal stress in the creep dynamics and failure of a biopolymer gel.

Francesco Piazza, Professor at University of Orleans, Center for Molecular Biophysics (CBM), UPR 4301 CNRS, University of Orleans.

Research Interests: Biological Physics, Statistical Physics, Molecular dynamics simulations.

Plenary talk title:

Ultra-coarse-grained models to simulate ligand-receptor kinetics of complex multivalent molecules: fundamentals and applications to the design of novel agents for immunotherapy.

IOP Conf. Series: Materials Science and Engineering 1160 (2021) 011001 doi:10.1088/1757-899X/1160/1/011001

Ashutosh Goel, Professor at the Department of Materials Science and Engineering, Rutgers The State University of New Jersey - USA.

Research Interests: *Glass/glass-ceramic, Optical Fiber technology, Bioactive glasses.*

Plenary talk title:

Turning Nuclear Waste to Glass in the United States: Past, Present, and Future.

Mohamed Naji, Professor at the Department of Physics, Laboratory of Applied Physics Informatics and Statistics, University of Sidi Mohamed Ben Abdellah, Morocco.

Research Interests: *Glass/ glass-ceramic, Ionic conductive glass, Relaxation, Glass dynamics.*

Plenary talk title:

Structure and relaxation dynamics of calcium-silicate glasses: Multi-scale approach

IOP Conf. Series: Materials Science and Engineering 1160 (2021) 011002 doi:10.1088/1757-899X/1160/1/011002

Peer review declaration

All conference organisers/editors are required to declare details about their peer review. Therefore, please provide the following information:

- **Type of peer review: Single-blind**

Single-anonymous: authors' identities are known to the reviewers, reviewers' identities are hidden from authors

- **Describe criteria used by Reviewers when accepting/declining papers. Was there the opportunity to resubmit articles after revisions?**

  Reviewers should consider the following key points related to scientific content, quality and presentation of the papers:

  - Technical Criteria
  - Scientific merit: notably scientific rigour, accuracy and correctness.
  - Clarity of expression; communication of ideas; readability and discussion of concepts.
  - Sufficient discussion of the context of the work, and suitable referencing.
  - Quality Criteria
  - Originality: Is the work relevant and novel?

- **Was there the opportunity to resubmit articles after revisions?**

  All revised articles were resubmitted with p-2-p letter after revision.

- **Conference submission management system:**
  Submission through the conference e-mail

- **Number of submissions received:** 13 papers

- **Number of submissions sent for review:** 12

- **Number of submissions accepted:** 11

- **Acceptance Rate (Number of Submissions Accepted / Number of Submissions Received X 100):** 84 %

- **Average number of reviews per paper:** 2

- **Total number of reviewers involved:** 14

- **Any additional info on review process (ie plagiarism check system):**

- **Contact person for queries:** mohamed.naji3@usmba.ac.ma

IOP Conf. Series: Materials Science and Engineering 1160 (2021) 012001 doi:10.1088/1757-899X/1160/1/012001

Structural and dielectric properties of Pb $(Zr_{0.52}Ti_{0.48})_{1-3x/4}Fe_x$ O_3 ceramics at $0\leq x\leq 0.20$ prepared by sol-gel method

Ahabboud Malika, Amarass M, Ahjyaje F Z, Abdi F and Lamcharfi T

Signals, Systems and Components Laboratory (LSSC), Electrical Engineering Department University Sidi Mohamed Ben Abdellah, Faculty of sciences and technologies. Fez, Imouzzer Road B.P. 2202, Morocco.
The corresponding author's e-mail: malikasmp2013@gmail.com (Ahabboud M).

Abstract. In this work, the Pb $(Zr_{0.52}Ti_{0.48})_{1-3x/4}Fe_x$ O_3 (x=0.0; 0.025; 0.05; 0.075; 0.10; 0.15; 0.20) ceramics were synthesized by using the sol-gel method. X-ray diffraction (XRD), Raman spectra, was used to reach the structural properties of the studied compounds. The XRD patterns registered at room temperature revealed the phase formation of the samples which was confirmed by the Rietveld refinement method which suggests the formation of mixed tetragonal and rhombohedral phases. Raman spectra show that when iron content increases, the intensity of Raman peaks of pure PZT increases. The dielectric properties of Pb $(Zr_{0.52}Ti_{0.48})_{1-3x/4}Fe_x$ (PZTFx) materials as a function of temperature are studied and displayed a resonance phenomenon for all samples which is moved to the higher frequencies combined with a decrease in dielectric constant when Fe concentration increase. The variation of the dielectric constant as a function of temperature at low frequencies reveals a shifting of the transition temperature to the lower temperature.

1. Introduction

Lead zirconate titanate solid solutions Pb $(Zr_{1-y}Ti_y)$ O_3 (PZT) is one of the most important members of piezoelectric ceramics because of their excellent properties. These materials are used in a large variety of uses including actuators and nonvolatile memory application, microelectromechanical systems, infrared detectors (MEMS), ultrasonic sensors, and non-volatile memories, photonic crystals, ferroelectric random access memory (FRAMs), terahertz emission, fluidic delivery, and nanosensors, etc [1-2]. The Pb $(Zr_{1-y}Ti_y)$ O_3 samples exhibit a phase diagram that is complex near the morphotropic phase boundary (MPB) at y=0.48 which shows high dielectric and excellent piezoelectric properties and the coexistence of two phases; rhombohedral (*R*3*m*) and tetragonal (*P*4*mm*) phases [3-5]. Depending on the exigency of the applications, commercially used PZT materials are enhanced by different dopants, and the general characteristics of $PbZr_{0.52}Ti_{0.48}O_3$ (PZT52) ceramic are generally improved by the addition of one cation which will replace the Pb^{2+} in site A or (Zr^{4+} / Ti^{4+}) in site B of the $PbZr_{0.52}Ti_{0.48}O_3$ (PZT52) perovskite. Here, we are concerned with the PZT ceramics which has been added Fe^{3+} into the PZT52 ceramic due to the presence of magnetic, and magnetoelectric coupling properties. These particles have been prepared by different methods such as chemical coprecipitation [6], mechanical alloying [7]. Hydrothermal method [3,8], conventional solid-state reaction [9-10], and sol-gel method [11]. However, in all of these and the preparation of PZT ceramics, problems typically encountered include loss of lead (Pb) due enormously to the volatility of PbO at elevated temperatures [2-3]. Now, in this work, we have displayed the synthesis of Fe doped PZT52 according to the formula (Pb $(Zr_{0.52}Ti_{0.48})_{1-3x/4}Fe_x$ O_3) samples using a sol-gel method. The X-ray diffraction techniques, Raman spectra, and dielectric properties as a function of temperature and as a function of the frequency of Fe doped PZT52 ceramics were investigated.

2. Experimental details

The PZTFx nanoparticles, where x=0 ;0.025; 0.05; 0.075; 0.10; 0.15 and 0.20, were prepared by the sol-gel method using a using Lead (II) acetate trihydrate [$PbC_2H_3O_2)_2.3H_2O$], Zirconium Acetate [Zr ($CH_3COO)_4$], titanium isopropoxide [Ti (OCH ($CH_3)_2)_4$], and Iron Acetate Fe (CH_3CO_2)$_2$ as starting precursors. The Metal acetates were dissolved in distilled water and mixed with stoichiometric amounts of sol [Ti (OCH ($CH_3)_2)_4$]. Then, the obtained solution was stirred for 1 hour to get a homogenous sol. After that, the sol was dried at 80 °C to remove water, and the PZTFx gel was formed. The resultant powder of PZTFx was calcined at 700 °C for 4 hours. The calcined powders were mixed with the polyvinyl alcohol PVA as a binder and pressed into pellets of diameter 12 mm and thickness approximately 1 mm. The pellets were placed in the rich PbO atmosphere using $PbTiO_3$ as a spacer powder to compensate for the loss of Pb probably evaporated during heat treatment [2]. These pellets were then sintered at 1100°C for 4 hours with a heating rate of 5 °C/min. The synthesized nanoparticles were characterized by X-ray diffraction (XRD) measurements with an (XPERT-PRO diffractometer system) with CuKα radiation with (λ=1.5406Å). Data were registered at room temperature in the angular range 20°-80° with steps of 0.04°. The lattice parameters were collected from the Rietveld method using Fullprof software of the experimental diffraction data. Raman spectroscopy was measured at room temperature. The dielectric measurements were investigated using an Agilent impedance analyzer (Agilent E4980A) in the frequency range from 100 Hz to 2 MHz and a wide range of temperatures (50–450 °C).

3. Results and discussion

Figure 1 a. illustrates the X-ray diffraction patterns of Pb ($Zr_{0.52}Ti_{0.48})_{1-3x/4}Fe_x$ O_3 (PZTFx) ceramics calcined at 700 °C for 4 hours. Powder XRD patterns indicate that the PZTFx crystallizes in the pure phase and there is no appearance of any impurity peaks. The zoom of the peak (110) is given in figure 1 b, it can be seen from this figure that the peak shifts to low angles for x=0.025 and high angles for 0.05≤ x≤0.20. This shifting and the change of intensity are probably due to the substitution of Zr^{4+} / Ti^{4+} in our opinion but it is also possible attributed to the substitution of Pb^{2+} by Fe^{3+}. These results agree with Gh.H. Khorrami and all of which studied the substitution of Ti^{4+} by Zn^{2+}, Mg^{2+}, and Mn^{2+} ions [12]. Rietveld refinement was carried out using the XRD data for PZTFx ceramics (x=0 ;0.025; 0.05; 0.075; 0.10; 0.15 and 0.20) shown in figure 2. The pseudo-Voigt function was used to define the peak profiles. Table.1 indicates the lattice parameters, the quality of fit (χ^2), and R-factors (R_p, R_{wp},) values calculated by Rietveld refinement as a function of iron concentration. As can be seen from the results of the refinement (Figure 2 and Table 1) that Rietveld refinement indicates a mixture of rhombohedral (R3m) and tetragonal (P4mm) crystalline phases for all samples with a percentage variation % phase. Indeed, from table 1, the Lattice parameters (a, c) are varied in fluctuation. This may be due to the compositional fluctuations, and substitutional disordering in the arrangement of cations in one or more crystallographic sites (Pb^{2+}, Zr^{4+}or Ti^{4+}) of the structure, this variation mode in lattice parameters is also found by J.A.Sanchez Caceres and al which they investigate the $PbZr_{0.52}Ti_{0.48}O_3$ compounds doped with erbium [13]. Moreover, The R_p, R_{wp}, values for all the samples are higher than 3,92, 4,97 respectively and the change of chi-squared (χ^2) is between 1.12 and 1.49 as a function of iron concentration. Additionally, the result of Rietveld refinement reveals for x= 0.05 is composed of 76.42%% tetragonal phase and 23.58% rhombohedral phase (Table 1), these results confirmed the existence of morphotropic phase boundaries in all the PZTFx ceramics.

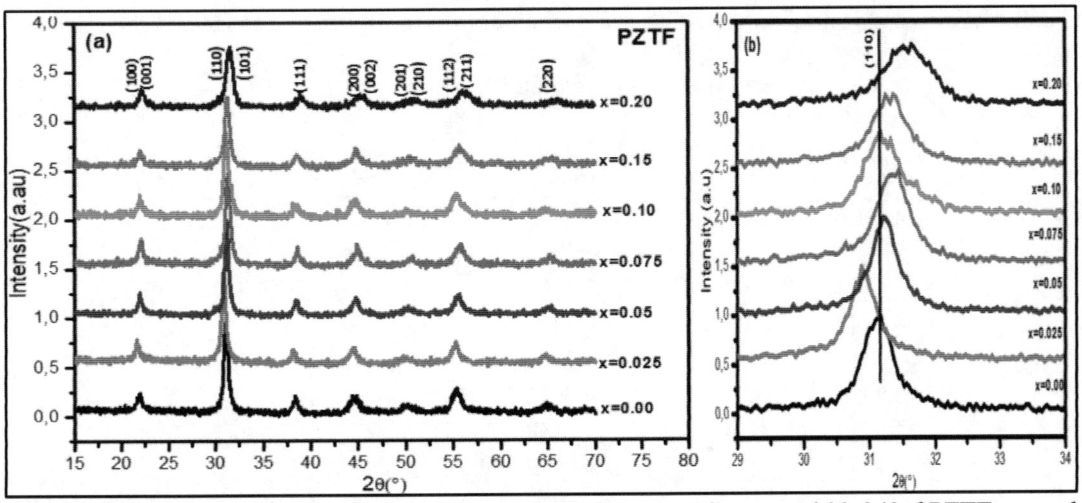

Figure 1 a. XRD patterns and b. Zoom of the peak (110) in the 2θ range of 29-34°of PZTFx samples calcined at 700 °C for 4h.

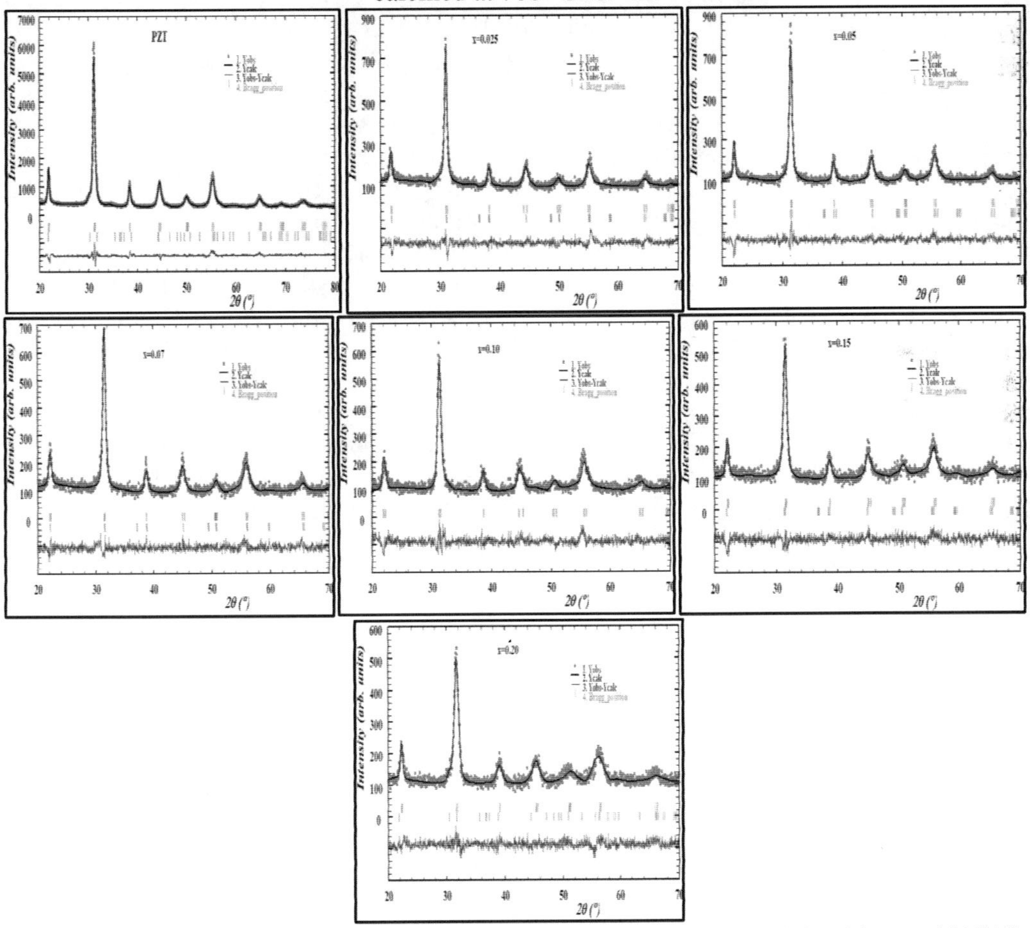

Figure 2. Rietveld refined the XRD pattern for the PZTFx samples calcined for at 700°C/4h.

Table 1. Lattice parameters(a,c),Rp, R_{W-P},and χ^2, values for the structures of PZTFx ceramics for $0 \leq x \leq 0.20$.

x	Lattice parameters (a,c) and % phases				Rp (%)	R_{wp}	χ^2
	Rhom		Tet				
	(a,c)	%	(a,c)	%			
X=0.00	a=5.7510 c=14.1262	48.42%	a=4.0402 c=4.1098	51.58%	3,92	4,97	1,12
X=0.025	a=5.7725 c=14.1946	4.59%	a=4.0736 c=4.1194	95.41%	8.38	10.6	1.43
X=0.05	a=5.7300 c=13.8989	23.58%	a=4.0324 c=4.0610	76.42%	7.82	9.95	1.30
X=0.075	a=5.6939 c=13.9257	18.94%	a=4.0370 c=4.0059	81.06%	7.51	9.47	1.14
X=0.10	a=5.7216 c=13.7214	2.66%	a=4.0626 c=4.0056	97.34%	8.70	10.9	1.41
X=0.15	a=5.7370 c=13.9696	14.13%	a=4.0417 c=4.0077	85.87%	8.52	10.8	1.49
X=0.20	a=5.6554 c=14.7054	2.32%	a=4.0026 c=3.9816	97.68%	7.89	10.1	1.32

Raman spectra of the PZTFx samples for compositions (x=0 ;0.025; 0.05; 0.075; 0.10; 0.15 and 0.20) at room temperature effected in the in the wavenumber range 100–1000 cm^{-1} are presented in figure 3. These peaks are designated based on the previous reports [14–16]. For x = 0.00 compound, we have observed eight main peaks: 203; 274; 331; 507; and 552; 597; 717; and 775cm^{-1}, corresponding to the E(2TO), E(silent) + B1, A1(2TO), E(3TO), E(4TO), A1(3TO), E(4LO), and A1(3LO) modes [14], respectively. The E+B1 (silent) mode confirms the existence of rhombohedral and tetragonal structures [17]. The modes below 340 cm^{-1} may correspond to the distortion of the BO6 octahedral. Modes in the range 340–598 cm^{-1} correspond to the O–B–O bending and 700–775 cm1 correspond to B–O stretching. These results are in good agreement with those reported in the literature [15]. similar peaks of pure PZT52 were observed in all the PZTFx samples furthermore, the intensity of these Raman peaks has not only increased but also shifted to the lower wavenumbers and this is well observed from the composition of x=0. 025.only and when the Fe increases, the intensities at 331cm^{-1} decrease and disappears completely for concentration higher than x =0.075. These results reveal that the replacement of (Zr, Ti) by iron in the B site of the crystal structure, as a result, reduces the B - O binding energy with increasing x content from 7.5%.

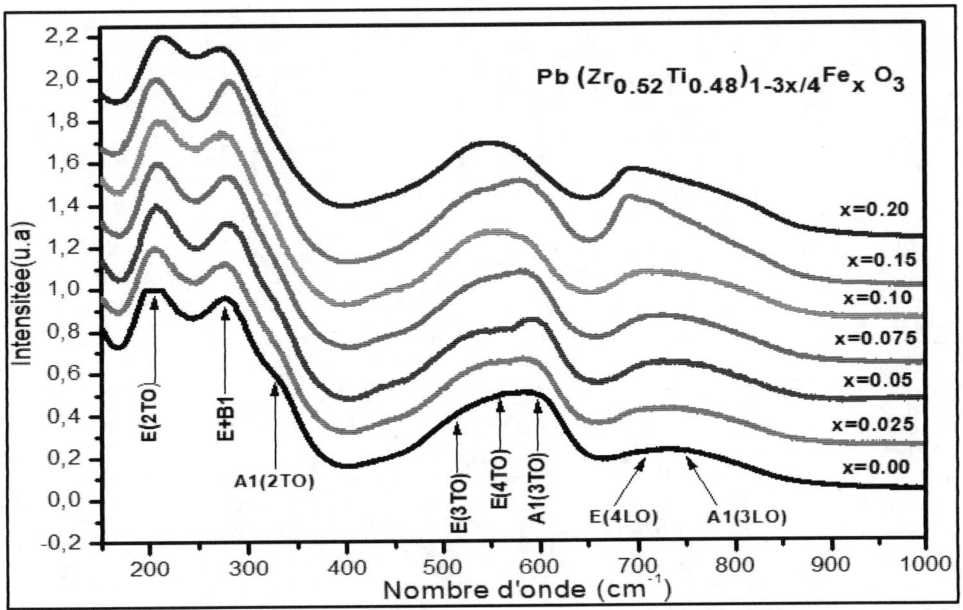

Figure 3. Raman spectra of Pb $(Zr_{0.52}Ti_{0.48})_{1-3x/4}Fe_x$ O_3 ceramics with different compositions.

The dielectric constant of PZTFx ceramics with x=0 ;0.025; 0.05; 0.075; 0.10; 0.15 and 0.20 were checked at different frequencies (low frequency f<1MHz) as a function of temperatures in the range of 50 to 400C°, and are shown in figure 4. It clear from this figure that for the undoped sample (PZT52) that the transition temperature from the ferroelectric phase to the paraelectric phase [8], corresponding to the maximum of the dielectric constant at the temperature T_c=410 °C named Curie temperature. The same evolution of dielectric permittivity as a function of temperature is observed in Fe doped PZT samples. The transition is found to shift to the lower temperature when the Fe concentration increases, indicating the effect of Fe doping on dielectric proprieties of PZT52 ceramics. These results indicate that Fe dopant might induce a diffused phase transition, resulting in the order-disorder phenomenon. Consequently, the dielectric constant maximum value of the PZTFx ceramics decreases with the increase of Fe substitution comparing to pure PZT. At 50 kHz the value is around 5652 for the PZTF5 ceramic compared to 18455 for the PZT52 ceramic. The obtained results are consistent with the results reported in the literature [4]. Figure 5 represents the evolution of constant dielectric as a function of temperature at different frequencies (at higher frequencies f>1MHz) for studied PZTFx ceramics. It is observed, for all the samples, that The constant dielectric increases with the increase of temperature and reaches a maximum value at Tm (maximum temperature) followed by a decrease. when the frequency increases, the maximum dielectric constant ε_{max} decreases, and the temperature Tm is found to shift to the lower temperature near the transition temperature Tc. This behavior is explained by the phenomenon of dielectric resonance [18-19]. It's also noted from this figure that when the x content increases, the permittivity maximum value ε_{max}, and T_m are found to decrease progressively compared to pure PZT52. For x=0.00, The ε_{max} is about 5803 and 3698 at 1MHz and 1.8 MHz respectively. For x=0.075, the dielectric constant maximum ε_{max} is around 2509 and 2357 at 1MHz and 1.8 MHz respectively.

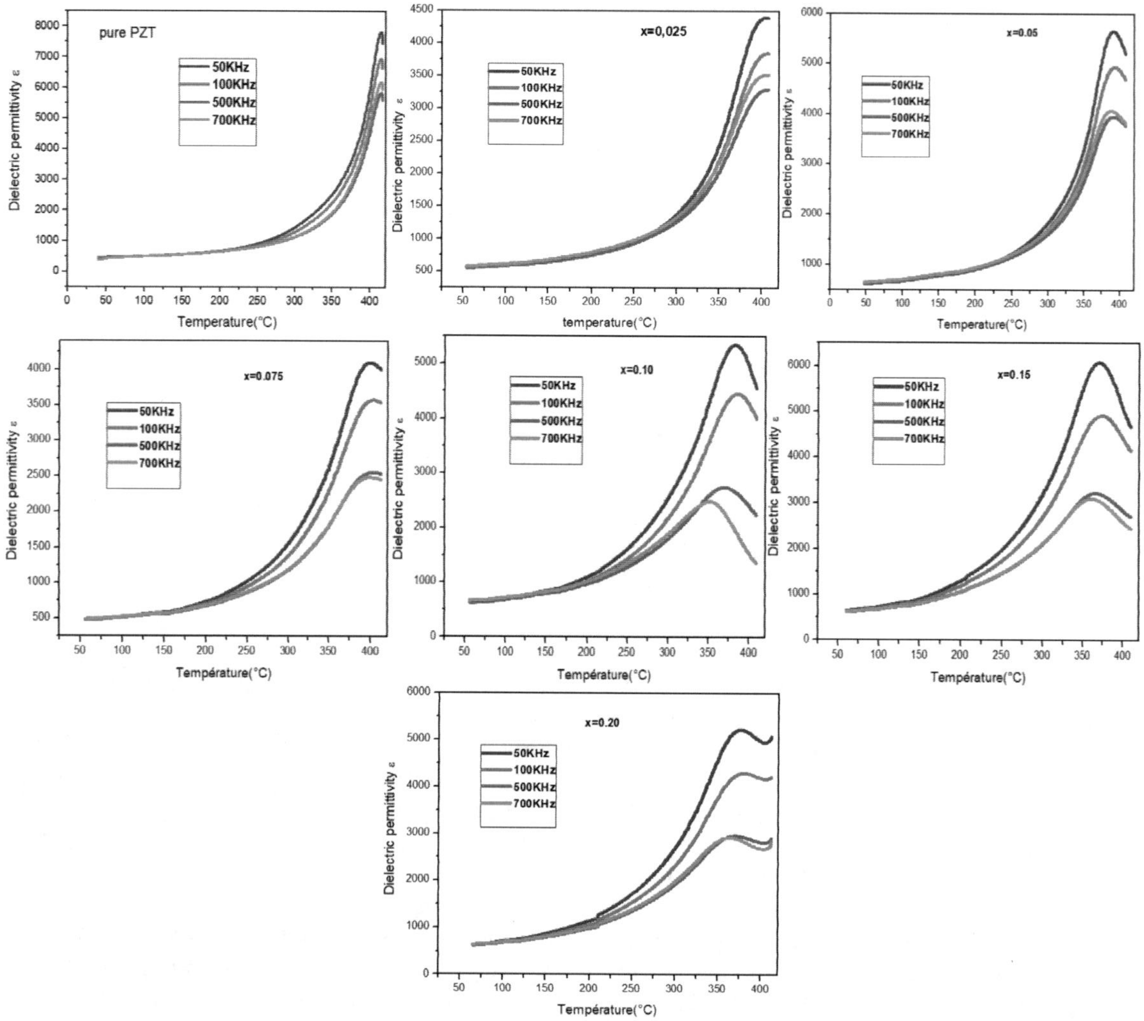

Figure 4. The temperature-dependent dielectric constant of PZTF ceramics at different frequencies f<1MHz.

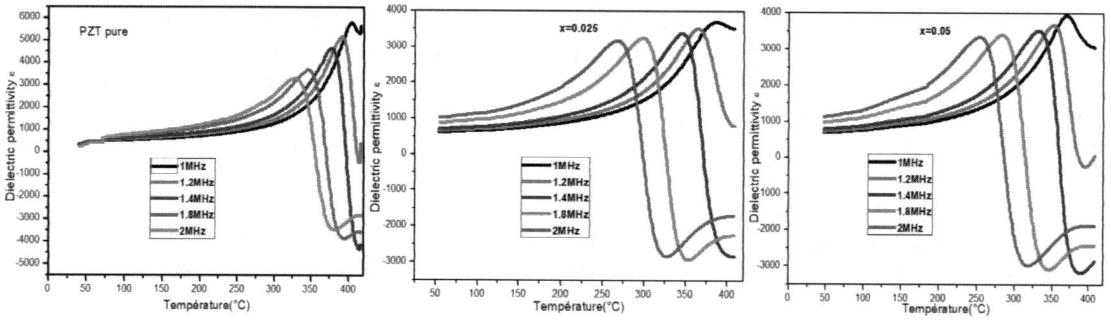

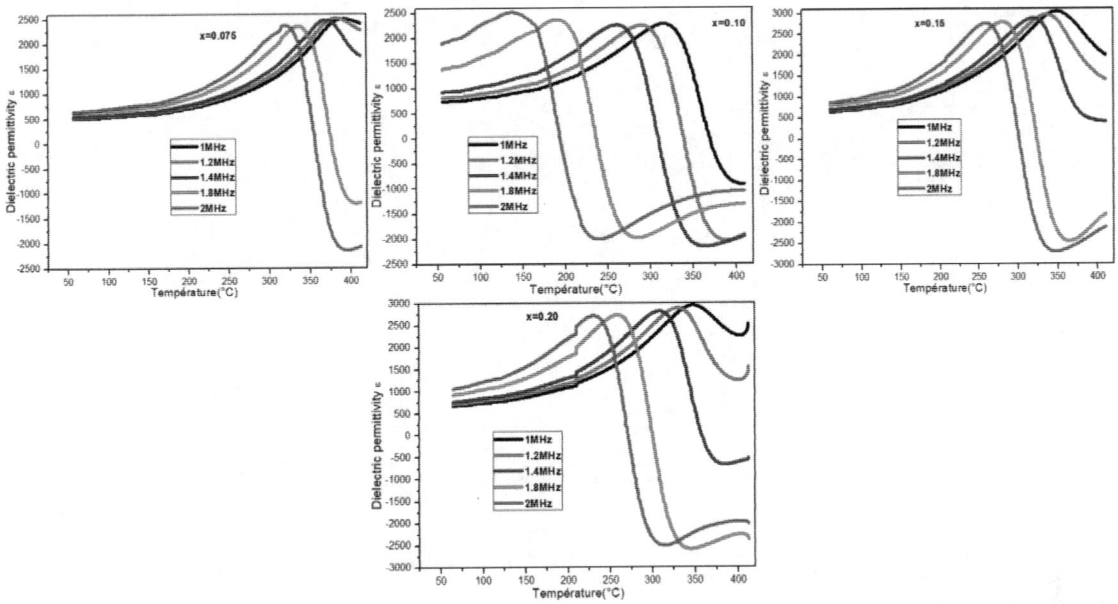

Figure 5. The temperature-dependent dielectric constant of PZTF ceramics at different frequencies f>1MHz.

The real part of the dielectric constant as a function of frequency and selected temperature (160 and 400°C) for PZTFx ceramics was determined and is presented in figure 6. The results of dielectric measurements for the PZT52 compound reveals that the dielectric permittivity increases with the increase of frequency and goes through a maximum value ε_{rmax} at the resonance frequency (maximum value) which displaced towards lower frequencies when the temperature increases. From this maximum, the spectrum of ε_r decreases and, go through a minimum at a particular frequency named anti-resonant frequency. This variation of real permittivity is explained by the phenomenon of dielectric resonance [19]. The dielectric resonance is found for all the prepared PZTFx nanoparticles and the dielectric constant maximum value decreases as a function of Fe concentration. In addition to this, the value of resonance frequency is found to increase with increasing Fe concentration. The Fe substitution decreases the dielectric constant of the samples.

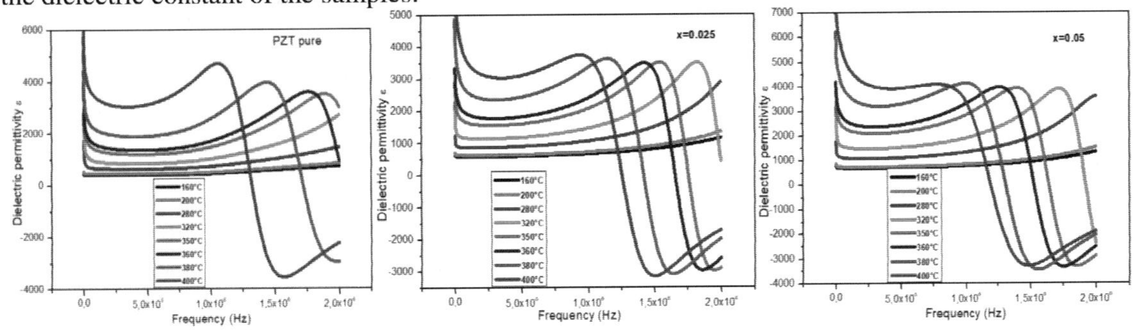

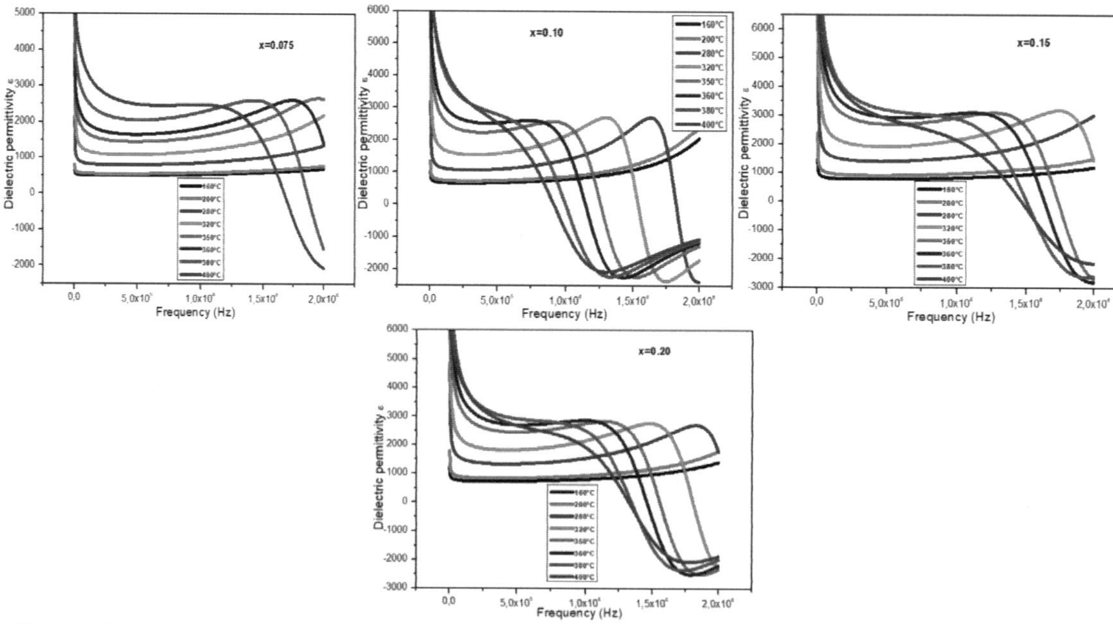

Figure 6. The frequency-dependent dielectric constant of PC_xZT ceramics at different measurement temperatures.

4.Conclusion

In summary, PZTFx (x=0;0.025; 0.05; 0.075; 0.10; 0.15 and 0.20) nanoparticles were synthesized using fine powders obtained by sol-gel method and characterized by XRD, Raman spectroscopy, and the dielectric properties as a function of temperature and frequency. The XRD data confirm the formation of the crystalline structures without any impurity peaks. Rietveld refinement method indicates the existence of MPB by the coexistence of rhombohedral (R3m) and tetragonal (P4mm) crystalline phases for all samples PZTFx, with a percentage variation % phase. It revealed also that the lattice parameters(a,c) are varied in fluctuation which probably due to substitutional disordering in the arrangement of cations crystallographic sites (Pb^{2+}, Zr^{4+}or Ti^{4+}) of the structure. The Raman spectroscopy shows the increase in intensity and disappearance of the A1(2TO) peak observed in the pure PZT52 material, for a percentage higher than x =0.075. Dielectric measurements showed a shifting of the phase transition to the lower temperature and a decrease in the dielectric constant maximum when the iron content increased, Also, the phenomenon of dielectric resonance are found at the higher frequencies f>1MHz in addition to the displacement of Tm to lower temperature when the Fe concentration increases. The evolution of the permittivity as a function of frequency shows that the value of resonance frequency increased with increasing Fe concentration.

5.References

[1] Suryakanta N, Tapan K C and Dipak K 2016 Spherical ferroelectric $PbZr_{0.52}Ti_{0.48}O_3$ nanoparticles with high permittivity: Switchable dielectric phase transition with temperature, *Journal of Ceramics International* vol 42 pp 14490–14498.

[2] Xiaoyan Z, et al 2016 Vertically Free-Standing Ordered Pb $(Zr_{0.52}Ti_{0.48})$ O_3 Nanocup Arrays by Template-Assisted Ion Beam Etching, *Nanoscale Research Letters SpringerOpen Journal* 11:225.

[3] Oliveiran C A, Longo E, Varela J A and Zaghete M A 2014 Synthesis and characterization of lead zirconate titanate (PZT) obtained by two chemical methods, *Journal of Ceramics International* vol 40 pp 1717–1722.

[4] Li J, 2011 Broadband Dielectric Response in Hard and Soft PZT: Understanding Softening and Hardening Mechanisms, *école Polytechnique fédérale de Lausanne* 4988.

[5] Subhash S, Vikash S, R K Kotnala, Rajeev R and R K Dwivedi, 2014 Co-existence of tetragonal and monoclinic phases and multiferroic properties for x=0.30 in the (1 x) Pb $(Zr_{0.52}Ti_{0.48})$ O_3–(x)$BiFeO_3$ system, *Journal of Alloys and Compounds* vol 614 pp 165–172.
[6] Avinandan M, Debasis G, Asish M, Parthajit P and Chapal K D 2013 Synthesis and Microwave Absorbing Properties of Cu-Doped Nickel Zinc Ferrite/Pb $(Zr_{0.52}Ti_{0.48})$ O_3 Nanocomposites, *Journal of Engineering*, Article ID 391083, p 8.
[7] Venkata R M et al 2010 Nanocrystalline Pb$(Zr_{0.52}Ti_{0.48})$ O_3 Ferroelectric Ceramics: Mechanical and Electrical Properties, *Journal of Nanomaterials*, Article ID 783043, p 8.
[8] Taj-dine L, 2005 Elaboration par voie hydrothermale et caractérisation des ceramiques ferroelectriques de type perovskite PZT dopées au lanthane Pb_{1-y} La_y (Zr_xTi_{1-x}) O_3, Thesis, *University of Bordeaux,France* .
[9] Venkata R M et al 2011 Investigation, and characterization of Pb $(Zr_{0.52}Ti_{0.48})$ O_3 nanocrystalline ferroelectric ceramics: By conventional and microwave sintering methods, *Materials Chemistry and Physics* Vol 126, 15 March 2011, p 295-300
[10] Noheda B, Gonzalo J A, 2000 A tetragonal-to-monoclinic phase transition in a ferroelectric perovskite: the structure of $PbZr_{0.52}Ti_{0.48}O_3$, *Physical review. B, Condensed matter The American Physical Society* vol 61, number 13.
[11] Kour P, Pawan K M K and Sinha S K, 2013 Enhance of ferroelectric properties by modifying Pb^{2+} side by Mg^{2+}in PZT (52/48) ceramics, *AIP Conference Proceedings* vol 1512, 1276.
[12] Khorrami Gh H, Khorsand Zak A, Kompany A and Ramin yousefi, 2012 Optical and structural properties of X-doped (X = Mn, Mg, and Zn) PZT nanoparticles by Kramers–Kronig and size strain plot methods, *Ceramics International* vol 38 pp 5683–5690.
[13] Jaime A S C, Carlos A C P, João V S C, Raoni C B and Rodolpho T C, 2019 Study of Structural and Electric Properties of the PZT 52/48 Doped with Er^{+3} *Materials Research.* Vol 22(5): e20190123
[14] Marco D, Hideo F, Nobuhiko T, Claudio C, Noriyuki H, Kenji K, Carmen G and Hiroshi H, 2011 Raman spectroscopic study of phase transitions in undoped morphotropic $PbZr_{1-x}Ti_xO_3$, *Journal of Raman Spectrosc* vol 42, pp 488–495.
[15] Souza F A G, Lima K C V, Ayala A P, Guedes I, Freire P T C, Melo F E A, Mendes F J, Araujo E B and Eiras J A, 2002 Raman scattering study of the $PbZr_{1-x}Ti_xO_3$ system: Rhombohedral-monoclinic-tetragonal phase transitions, *physical review* B 66, 132107.
[16] Rodrigues S A S, Rolo A G, Khodorov A, Pereira M and Gomes M J M, 2010 Determination of residual stress in PZT films produced by laser ablation with X-ray diffraction and Raman spectroscopy, *Journal of the European Ceramic Society* vol 30 pp 521–524.
[17] Paramjit K, Sudipta Kishore P, Pawan K, Sanjay Kumar S and Manoranjan K 2016 Enhanced Ferroeletric and Piezoelectric Properties of Nd^{3+} doped PZT nanoceramics: *AIP Conference Proceedings American Institute of Physics* vol 1728, 020490.
[18] Aimé P B and José D S G, 2010 Dielectric Relaxation Phenomenon in Ferroelectric Perovskite-related Structures, *Ferroelectrics Edited by Dr. Indrani Coondoo* ISBN 978-953-307-439-9 Hardcover, 450 pages.
[19] Mesrar M, Lamcharfi T, Echatoui N S, Abdi F, Ahjyaje F Z and Haddad M, 2019 Effect of barium doping on electrical and electromechanical properties of (1-x) $(Na_{0.5}Bi_{0.5})$ $TiO_{3-x}BaTiO_3$, *Mediterranean Journal of Chemistry*, 8(2), 198-20

Probing the balance of Telechelic Polymers Bridging and Screened-Coulomb Interactions in Microemulsion System

R Ahfir[1], S Elkhaoui[1], M Khatouri[1], M Lemaalem[2], L Talha[1]and M Filali[1]

[1] Laboratoire de Physique Appliquée, Informatique et Statistique (LPAIS), Faculty of Sciences Dhar El Mahraz, Sidi Mohamed Ben Abdellah University, BP 1796 Fès, Atlas, Morocco
[2] Laboratoire de Physique des Polymères et Phénomènes Critiques Sciences Faculty Ben M'Sik, Hassan II University, P. O. Box 7955, Casablanca, Morocco

* E-mail : mohammed.filali@usmba.ac.ma

Abstract. The structural properties of neutral oil-in-water (O/W) microemulsions nanodroplets in the presence of telechelic polymers PEO modified with two hydrophobic end functions (PEO-2m) were studied by Small-Angle Neutron Scattering (SANS). We show that the unknown attractive interaction introduced by the addition of PEO-2m, can could be balanced by a known repulsion interaction due to the addition of a cationic surfactant Cp^+, which causes a progressive transformation of the bridges connecting the nanodroplets into loops. For this, we compute the structure factor S(q) with an effective pair potential, using the Ornstein Zernicke (OZ) integral equation approach with the Hypernetted Chain (HNC) closure relation. Generally, the agreement between the neutron scattering spectra and the numerical study is reasonable and allows a detailed description for each sample.

1. Introduction

The microemulsions are a thermodynamically stable dispersion in water of oil nanodroplets surrounded by a surfactant film [1]. Microemulsions are widely used in a large array of chemical and biochemical systems including detergents, lubrificant, cosmetics, formulation of paints and advanced petrol extraction [2-6]. In recent decades, many studies have focused on mixtures between polymers and surfactant-water-oil systems [7-14]. These systems are of great practical importance, for example in the control of rheology and colloidal stability. Indeed, they serve as a model system for a general class of transient networks. The behavior of these mixed systems depends strongly on the interaction between the polymer and the surfactant film. Generally, for a given number of polymer chains per colloid, they have a monophasic region populated with flower-like colloids with a low droplet concentration, followed by a two-phase region due to attractive phase separation and finally a concentration phase region of connected colloids, with a gel-like behavior [15, 16]. The first theoretical approach was proposed by Milner and Witten, who adapted self-consistent field calculations to telechelic polymer brushes between flat plates [17], Meng and Russel [18], redid this calculation and they obtained a stronger attraction between the plates in better agreement with the experimental results [19, 20]. As regards the curved surfaces, the problem is more complicated, Semenov et al. have proposed a scaling approach [21]. The results for potentials published on telechelic chains connecting finite curvature surfaces come from Bhatia and Russel [22]. They obtained analytical results for plane plates and numerical results for chains between two spheres. Porte and al [23] used a model derived from Bhatia and Russel to analyze experimental spectra of small-angle neutron scattering of microemulsion nanodroplets connected by the telechelic polymers.

In this study, we plan to will describe and model interactions between spherical oil-in-water neutral microemulsions nanodroplets with telechelic polymers (PEO-2m), to which small amounts of cationic surfactant Cp^+ are added. Then, we use a microemulsion of decane stabilized in distilled water by a monolayer of non-ionic surfactant TX_{100} and co-surfactant TX_{35}. To this microemulsion, a telechelic polymer consisting of a water-soluble poly (ethylene oxide) chain, PEO (abbreviated), to which two hydrophobic ends $C_{12}H_{25}$ are grafted, it will be noted PEO-2m. Thus, the polymeric ends (stickers) are incorporated into the droplets to bridge them and form a multi-connected network. The presence of Cp^+

introduces a coulomb repulsive interaction. Our objective is to "offset" the unknown attractive interaction introduced by the addition of PEO-2m, by a known repulsion interaction. We want to model these interactions by an appropriate potential and to find the parameters of good agreement between the SANS experiments and the numerical study for all studied cases.

2. Experimental section

2.1. Materials and Preparation of the microemulsion droplets with polymers PEO-2m

Oil-in-Water microemulsion studied system contain a neutral surfactant TX_{100}, cosurfactant TX_{35}, oil of decane, and Cetylpyridinium Chloride (CpCl). All components are purchased from Fluca (cf. table 1 for details). Samples are prepared in bi-distilled water. The ratio in weight of TX_{35} to TX_{100} is set equal to 0.48. The ratio in weight of Decane to surfactant film is 0.76.

To the microemulsion, we add the hydrophobically modified poly (ethylene oxide) PEO-2m. PEO has been hydrophobically modified and purified in the laboratory using the method described in Refs. [24] and [25]. PEO-2m contains an isocyanate group between the alkyl chain $C_{12}H_{25}$ and the ethylene oxide chain. The samples are characterized by a volume fraction of droplets Φ , the number r of C_{12} chains per droplets. To obtain weakly charged nanodroplets, we substitute a small fraction of the non-ionic surfactant TX_{35} and TX_{100} with cationic surfactant. The charge rate τ is the ratio in weight of the cationic surfactant CpCl to the neutral surfactant film. The details on the components are indicated in Table 1.

Table 1. Molar mass and density of the components of the samples.

Components	Molar mass (Dalton)	Density (g/ cm^3)
H_2O	18	1
$(H_3C)_3C(CH_2)C(CH_3)_2C_6H_6(OCH_2CH_2)_{9.5}OH$ (TX100)	624	1.07
$(H_3C)_3C(CH_2)C(CH_3)_2C_6H_6(OCH_2CH_2)_3OH$ (TX35)	338	1.02
$[H_3C(CH_2)_8CH_3]$ (Decane)	142	0.73
$[H_3C(CH_2)_{15}]C_5H_5N^+Cl^-$ (CpCl)	339.5	1.656
$[CH_3(CH_2)_{11}NHCO(OCH_2CH_2)_{227}O(CO)NH\ [(CH_2)_{11}CH_3]$ (PEO-2m)	10400	1.2

2.2. Small-angle Neutron scattering (SANS) measurements

SANS experiments are carried out on the PACE spectrometer at the LLB-Saclay (France). We obtain intensities in absolute units (cm^{-1}) with accuracy better than 10%. To simulate correctly the experimental spectra, all the models spectra are convoluted by the instrumental response function taking into account the uncertainty of the neutron wavelength and angular definition [26]. The scattered SANS intensity I(q) by a dispersion of spherical colloids with a moderately spread distribution of size was modeled using the following expression [27]:

$$I(q)=\frac{1}{V}<A(\vec{q}).A^*(\vec{q})>=\Phi.v.(\Delta\rho)^2P(q)S(q) \quad (1)$$

where $q(\text{Å}^{-1})$ is the scattering vector, Φ is the volume fraction of the aggregates, $v(cm^3)$ is the dry volume of the aggregates, $\Delta\rho=6.6\times10^{10}cm^{-2}$ is the contrast, i.e., the difference in the scattering length density of the aggregates and of the solvent. P(q) is the form factor of the colloidal aggregates. The analysis of measurable structure factor S(q) allows to extract information on about the interdroplets potential of the system; i.e., the interactions between nanodroplets are repulsive or attractive.

3. Theoretical background

3.1 Expression of the pair potential

The total interaction potential between droplets with PEO-2m and the surfactant Cp^+ is written:

$$U_{tot}(r)=U_{HS}(r)+U_{VDW}(r)+W(r)+V(r)+U_{Coulomb} \quad (2)$$

- The simplest potential of pair which is between hard spheres of diameter 2R is:

$$U_{HS}(r)=\begin{cases}\infty & \text{for:} \quad r\leq 2R\\ 0 & \text{for:} \quad r>2R\end{cases} \tag{3}$$

Here, ‘r’ is the center-to-center distance; R is the radius of the nanodroplets.

- The attractive interaction of van der Waals acts at distances between nanodroplets of the order of 10% of their diameter. In the case of two dispersed spheres, U_{vdW} potential [28] :

$$U_{VdW}(r)=-\frac{A_H}{6}\left(\frac{2R^2}{r^2-4R^2}+\frac{2R^2}{r^2}+\ln\left(\frac{r^2-4R^2}{r^2}\right)\right),\quad \text{for:}\quad r>2R \tag{4}$$

A_H is the effective Hamaker constant, here, we consider $A_H=1.1k_BT$, this value is appropriate for decane nanodroplets interacting through water [29]. To the neutral microemulsion, we add telechelic polymers with two end (PEO-2m), an additional attractive and a steric repulsive contribution must be taken into account:

$$V(r)=V_{steric}\left(\frac{2R}{r}\right)\exp\left(-\frac{r-2R}{\lambda_{steric}}\right) \tag{5}$$

and

$$W(r)=W_{bridging}\left(\frac{2R}{r}\right)\exp\left(-\frac{r-2R}{\lambda_{bridging}}\right) \tag{6}$$

where V_{steric} and $W_{bridging}$ are the contact potential for the steric and the bridging interaction. The range of interactions of the polymers is set by their size. The parameters have been fixed to an identical interaction range λ_{steric} of 35Å and $\lambda_{bridging}$ of 100Å.

- When the cationic surfactant Cp^+ is added to the system (microemulsion with PEO-2m), a screened Coulombic repulsive potential [30, 31] is added to the existing interaction:

$$U_{Coulomb}(r)=k_BT\frac{Z_{eff}^2 l_B}{(1+\kappa R)^2}\frac{e^{-\kappa(r-2R)}}{r} \tag{7}$$

The range of repulsive interactions is determined by the Debye length k^{-1}, with a Bjerrum length of $l_B=\frac{e^2}{4\pi\varepsilon_0\varepsilon_r k_BT}=7.18\text{Å}$ and $T = 298K$. Z_{eff} is the number of charge effective per nanodroplets [32].

In the calculation of $I(q)$, Z_{eff} is the only adjustable parameter.

The total dimensionless pair potential, $\frac{U(r)}{k_BT}$ of expression (2) is:

$$\frac{U(r)}{k_BT}=\begin{cases}\infty & \text{for : } r\leq 2R\\ -\frac{1.1}{6}\left(\frac{2R^2}{r^2-4R^2}+\frac{2R^2}{r^2}+\ln\left(\frac{r^2-4R^2}{r^2}\right)\right)+\frac{Z_{eff}^2 l_B}{(1+\kappa R)^2}\frac{e^{-\kappa(r-2R)}}{r} & \\ +\frac{2R.V_{steric}}{k_BT}\frac{e^{-(r-2R)/\lambda_{steric}}}{r}+\frac{2R.W_{bridging}}{k_BT}\frac{e^{-(r-2R)/\lambda_{bridging}}}{r} & \text{for : } r>2R\end{cases}$$

3.2 Integral Equation Method

The Ornstein–Zernike (OZ) integral equation is written:

$$h(\vec{r})=c(\vec{r})+\rho\int c(|\vec{r}-\vec{r}'|)\,h(\vec{r}')\,d\vec{r} \tag{8}$$

The (OZ) equation related total; $h(\vec{r})=g(\vec{r})-1$ and direct $c(\vec{r})$ correlation functions of two particles at distance ‘r’ apart [33, 34]. To compute $g(r)$, the (OZ) equation was solved using the hypernetted-chain approximation (HNC) [35].

In terms of c(r), the (HNC) closure relation is: $c(r)\approx -\beta U(r)+h(r)-\ln[1+h(r)]$ (9)

IOP Conf. Series: Materials Science and Engineering 1160 (2021) 012002 doi:10.1088/1757-899X/1160/1/012002

Exponentiation gives: $$g(r) \approx \exp\left[-\beta U(r) + h(r) - c(r)\right] \quad (10)$$

For an isotropic colloidal fluid, the information on the radially averaged near-range ordering, i.e. the conditional probability of finding the center of colloid nanodroplets at separation 'r' from another one [36] is encoded in the equilibrium RDF g(r). In the reciprocal wavenumber space, the same structural information is included in the structure factor S(q).

4. Results and discussion

4.1 Phase behavior

The phase behavior of the quaternary system (TX_{100}, TX_{35}, decane, water) has been detailed in the literature [37-39]. For a small amount of PEO-2m $(r < 12m)$, the solution is mono-phasic separated by a percolation line between a sol phase and a gel phase [40]. For the same volume fraction $(\Phi < 7\%)$ and by increasing the amount of PEO-2m $(r > 12m)$, one observes an associative phase separation between a concentrated and very diluted a solution [41]. In this study, we investigate samples with $\Phi = 2.8\%$ and $r = 29.6\text{-}2m$. Thus, in this studied composition, the sample undergoes phase separation and is biphasic. The addition of a POE-2m polymer $(r = 29.6)$ on a microemulsion will cause an associative phase separation due to the effective attractive interaction between droplets in the low volume fraction. To avoid this phase separation, an adequate amount of cationic surfactant CpCl must be added to the solution. The latter introduces a repulsive interaction (electrostatic effect) which will be compensated by the attractive interaction due to the bridging. Thus, a single-phase solution is obtained. Our particular interest is to " weight " this unknown attractive interaction by a known repulsive interaction.

4.2 Nanodroplets shape and size

From SANS experiments, many studies have been carried out on different systems (bare microemulsion, microemulsion with POE-2m and / or the cationic surfactant (Cp^{+})) and have shown that the spherical shape and the size of the microemulsion droplets remain identical from Φ=1.4% to 26.5%, with an average radius $\overline{R}$=82Å and a polydispersity ΔR=15Å [12,23,43].

4.3 Structure Factor of mixture system (microemulsion + PEO-2m + Cp^{+})

In Figures. 1, 2 and 3, the SANS pattern of the bare microemulsion $(r = 0)$ is compared to that of a microemulsion with PEO-2m $(r = 29.6\text{-}2m)$ at different charges rates τ (+0.22%, +0.5%, +0.75% and +1%).

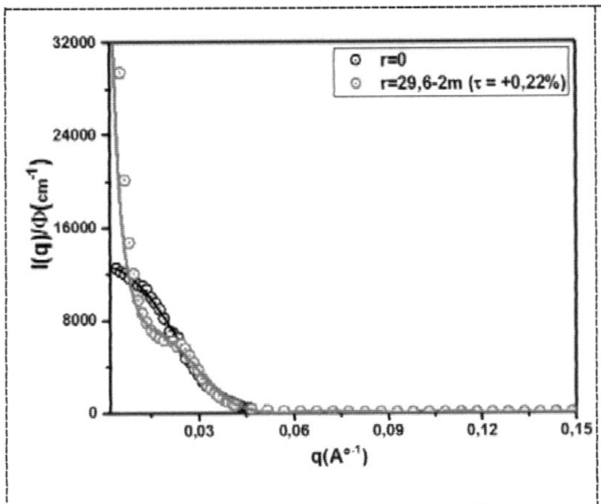

Figure 1. Scattered intensity I (q) from SANS and HNC at Φ=2.8%, for bare microemulsion (○ r=0) and with PEO-2m $(r = 29.6)$, at τ =+0.22% (○)

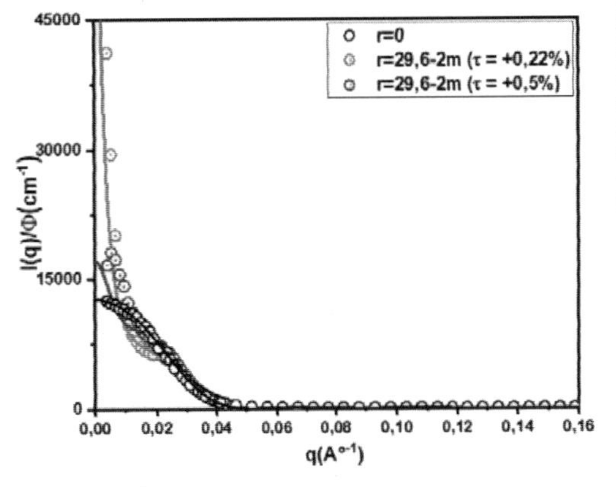

Figure 2. Scattered intensity I (q) from SANS and HNC at Φ=2.8%, for bare microemulsion (○r=0) and with PEO-2m $(r = 29.6)$, at τ =+0.22% (○) and τ =+0.5% (○)

For a low charge rate τ =+0.22% (Figure. 1), a significant rise in the scattered intensity is observed for $q \to 0$ (the osmotic compressibility) and the presence of a correlation bump of a particular distance between the droplets is favored ($d \sim 332Å$). This distance is imposed by the average end-to-end distance of the telechelic polymers [13]. Furthermore, it's worth noting that the significant increase in intensity is reasonable because the repulsive effect introduced by the charges (Cp+) is very weak compared to the still dominant attractive effect brought by PEO-2m.The best fit of the curve of the sample, using the model described by eq 2 yields an average number of charges per droplet $\bar{p}$ = 8(Z_{eff} =4.8) and a bridging potential $W_{bridging} = -4.02k_BT$.

Figure 2 displays the variation of the scattered intensity with q for charges rates $\tau = 0$, +0.22% and +0.5%. The scattered intensity decreases at low q when the amount of added charges is increased from τ = +0.22% to +0.5%, demonstrating that the repulsive effect is reinforced. This is illustrated by the decrease of $I(0)$ from 43000 cm^{-1} to 16000 cm^{-1}. The shoulder gives a particular distance between the droplets ($d \sim 353Å$).

Furthermore, an excellent quantitative agreement is found between the experimental and calculated spectra with average number of charges per droplet $\bar{p}$ = 19.15 (Z_{eff} =8) and a bridging potential $W_{bridging} = -3.9k_BT$.

For higher charges rates, τ = +0.75% and +1% (Figure. 3), we note the appearance of a very well defined sharp correlation peak for these two samples. In addition, the maximum of the correlation peak clearly shifts to large q's as τ increases. Then, the distance between the droplets decreases and the solution becomes more structured.

For τ = +0.75%, the best fit is obtained with an average number of charges per droplet $\bar{p}$ = 28.6 (Z_{eff} =17) and a bridging potential $W_{bridging} = -2.8k_BT$. For τ = +1%, $\bar{p}$ = 38 (Z_{eff} =21), $W_{bridging} = -2.5k_BT$.

Regarding the scattered intensity at low q, it is observed to decrease when the charge rate τ increases from τ = +0.75% to τ = +1%. This is a signature of an increasing correlation between the nanodroplets and consequently the repulsive interaction is more reinforced. This evolution shows the attractive interaction introduced by POE-2m is gradually compensated and replaced by Coulomb repulsion. As a result, we can probe the balance of telechelic polymers bridging and Screened-Coulomb Interactions for such systems. Alternatively, we can use a known repulsive interaction (Coulomb interaction) to "weight" a not well-defined attractive interaction (attraction by bridging).

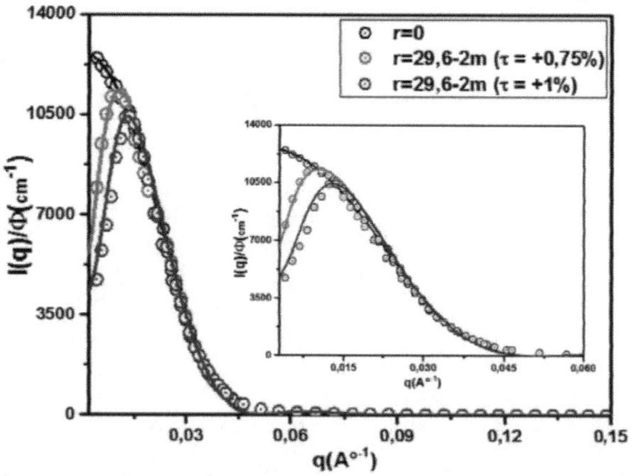

Figure 3. Scattered intensity I (q) from SANS and HNC at Φ=2.8% , for bare microemulsion (○ r=0) and with PEO-2m (r = 29.6), at τ =+0.75% (○) and τ =+1% (○)

We now examine the evolution of these samples from their phase behavior when the charge rate τ is increased. For low charge rates (+0.22%, 0.5%), the attractive effect is always present and it is reminiscent of the existence of some bridges between the droplets. For the large values of τ (+0.75%, +1%), the attractive effect no longer exists and the solution is completely fluid. This means that the nanodroplets are not connected between them and are instead decorated by loops (the two hydrophobic ends of the polymers stick in the same nanodroplet). As a result, there is a gradual transformation of the bridges connecting the neighbouring nanodroplets into loops. In Table 2, we report all results found for samples studied.

Table 2. Comparison of the results for $\Phi = 2.8\%$ of the distance between droplets (d), their effective charges Z_{eff}, $I(0)$, the repulsive potential in contact U_{0C}, bridging potential ($W_{bridging}$), and steric potential (V_{steric}) for the microemulsion with PEO-2m ($r = 29.6$), at different charge rates of Cp^+ τ (+0.22%, +0.5%, +0.75% and +1%)

	With POE-2m+charges					
	I(0) (cm^{-1})	d (Å)	Z_{eff}	$U_{OC}(K_BT)$	$V_{steric}(K_BT)$	$W_{bridging}(K_BT)$
$\tau = +0.22\%$	29436	332	4.8	0.37	10.5	-4.02
$\tau = +0.5\%$	16640	353	8	1.05	10.5	-3.9
$\tau = +0.75\%$	7938	409	17	4.75	10.5	-2.8
$\tau = +1\%$	4733	487	21	7.25	10.5	-2.5

5. Conclusion

In this paper, we use small angle neutron scattering to "weight" an unknown effective attraction against using a well-known electrostatic repulsion. The scattered intensity can be reproduced qualitatively using the (OZ) equation, with the (HNC) approximation. The interactions induced by the presence of PEO-2m and Cp^+ are modeled by Yukawa type potentials. The quantification of the attractive interaction is made by determining the bridging potential in contact $W_{bridging}$. In each case, we found the effective charge per droplet Z_{eff} and the Coulombic potential at contact U_{0C}. The strongly increase of the electrostatic repulsion strongly depends on the number of charges per droplet and for the large values of τ, the system is fluid. We obtained a reasonable adjustments for all cases studied. Our results indicate good matching of the parameters of this potential model.

Acknowledgments

This work was achieved in collaboration with the Laboratory of Montpellier (Charles Coulomb). We are much indebted to Professors G. Porte and J. Appell, for their assistance and their support during the realization for this work, R. Aznar, for synthesizing the copolymer used in this work. The SANS experiments have been performed on line PACE at Laboratory Leon Brillouin–CEA–CNRS. We thank Loic Auvray and Didier Lairez for their help. We thank heartily Luc Belloni who gave us the code for the computation of the structure factor, and for the fruitful discussions which enabled us the determination of effective charge.

6. References

[1] Schulman J H, Stoeckenius W 1959 *J. Phys. Chem.* **63** 1677
[2] Du Z, Mao X, Tai X, Wang G and Liu X 2016 *Journal of Molecular Liquids*. **223**, 805-810
[3] Dong B, Zhang Y, Qiu Z, Lu Z, Liu J, Pan L and He P 2018 *International Petroleum and Petrochemical Technology Conference* (pp. 29-37). Springer, Singapore
[4] Volpe V, Nascimento D S, Insausti M and Grünhut M 2018 *Colloids and Surfaces A: Physicochemical and Engineering Aspects*. **546**, 285-292
[5] Ormsby B, et al. 2016 *Journal of the American Institute for Conservation*, **55**(1), 12-31
[6] Degot P, et al. 2020 *Food Chemistry*, **336**, 127-660
[7] Kabalnov A, Olsson U, Wennerstroem H 1994 *Langmuir*. **10**, 7, 2159-2169

[8] Kabalnov A, Olsson U, Thuresson K, Wennerstroem H 1994 *Langmuir.* **10**, 12, 4509-4513
[9] Clegg S M, Williams P A, P Warren and Robb I D 1994 *Langmuir.* **10**, 3390
[10] Bagger-Jörgensen H, Olsson U, Iliopoulos I 1995 *Langmuir.* **11**, 6, 1934−1941
[11] Ahfir R, Azougarh A, Benhamou M and Filali M 2014 *International Review of Physics* **8** 3
[12] Elghazrani K, Azougarh A, Oberdisse J and Filali M 2014 *Eur. Phys.* J. E **37**, 1−10
[13] AHFIR R, Filali M 2020 *Adv. Mater. Lett.* **11** (8), 20081548
[14] Lemaalam M, AHFIR R, Derouiche A and Filali F 2020 *RSC Adv.* **10**, 36155
[15] Annable T, Buscall R, Ettelaie R, Whittlestone D and Rheol J 1993. **37**, 695
[16] Odenwald M, Eicke H F and Meier W 1995 *Macromolecules* **28**, 5069
[17] Milner S T, Witten T A 1992 **25** (20), 5495–5503
[18] Meng X X, Russel W B 2003 *Macromolecules* **36** (26), 10112–10119
[19] Eiser E, Klein J, Witten T A, Fetters L J 1999 *Phys Rev Letts.* **82** (25), 5076–5079
[20] Dai L, Toprakcioglu C 1991 *Euro phys Lett.* **16** (4), 331–335
[21] Semenov A N, Joanny J F, Khokhlov A R 1995 *Macromolecules.* **28** (4), 1066–1075
[22] Bhatia S R, Russel W B 2000 *Macromolecules.* **33** (15), 5713–5720
[23] Porte G, Ligoure C, Appell J, Aznar R 2006 *J Stat Mech Theory Exp.* **05**, 05005
[24] Kaczmarski J P, Glass J E 1993 *Macromolecules.* **26**, 5149
[25] Vorobyova O, Yecta A, Winnik M A, Lau W 1998 *Macromolecules.* **31**, 1, 8998-9007
[26] Lairez D 1999 *J Phys IV Fr.* **9**, 1-67
[27] Grillo I 2008 Small-angle neutron scattering and applications in soft condensed matter in Borsali R, Pecora R(Eds.), Soft Matter Characterization, Springer Netherlands, 723-782.
[28] Russel W B and Schowalter W R 1989 Colloidal Dispersions. Cambridge University Press
[29] Israelachvili J 1992 Intermolecular and Surfaces Forces. New York Academic, p228
[30] Belloni L 2000 Colloidal interactions *J. Phys Condens. Matter* **12** R549
[31] L. Belloni, Belloni L 2002 *J PhysvCondens Matter* **14** 9323
[32] Alexander S, Chaikin P M, Grant P, Pincus P and Hone D 1980 *J. Chem. Phys.* **80**, 5776
[33] Ornstein L.S and Zernike F 1914 *Proc Akad* Sci **17**, 793
[34] Hansen J P and McDonald I R 1986 Theory of Simple Liquids 2nd ed. Academic, London
[35] Van Leuwen J M J, Groenveld J and De Boer J 1959 *Physica* Amsterdam **25**, 792
[36] Banchio A J and Nägele G 2008 *Journal of chemical physics* **128**, 104903
[37] Filali M, Aznar R, Svenson M, Porte G and Appell J 1999 *J Phys Chem B* **103** 7293
[38] Filali M, Ouazzani M J, Michel E and G. Porte 2001 *J. Phys. Chem. B* **105**, 10528
[39] Porte G, Berret J F and Harden J L 1997 *J Physique II* **7**, 459
[40] Michel E, Filali M, Aznar R, Porte G, Appell J 2000 *Langmuir* **16**, 8702
[41] Piculell L and Lindman B 1992 *Adv. Colloid Interface* Sci. **41** 149
[42] Puech N, Mora S, Testard V and Oberdisse J 2008 *Eur. Phys. J. E Soft Matter 26* **13**, 2008.
[43] Talha L, Ahfir R, Azougarh A and Filali M 2015 Review of Physics **8** 4

An approach to determine the diatomic vibrational energy levels: application to LiH, NO and CO molecules

I Amila*, M Janati Idrissi, A Fedoul* and S Sayouri
Laboratory of Applied Physics Informatics and Statistics
BP 1796 Fes-Atlas, Morocco
**Corresponding authors: ibrahimamila@gmail.com or afedoul@yahoo.fr*

Abstract. This work presents the systematic calculation of vibrational energy levels, for diatomic molecules through an alternative approach, reposed on the Floquet representation analysis. The Morse potential function has been employed to solve the Schrodinger equation, and to obtain the vibrational energy levels of diatomic systems. As an illustration, the numerical energy values of some molecules have been computed, and compared to those of literature, revealing a good agreement, and showing the accuracy of our approach.

Keywords: Energy level; Diatomic; Floquet representation; Morse potential.

1. Introduction

In this work, we are interested in solving Schrödinger equation for Morse potential, which is a realistic model to describe the vibrational energy levels of diatomic molecules, with an appropriate approximation. Several analytical and numerical methods have been developed and applied to provide the resolution of this problem among them: the popular perturbation theory [1], Nikiforv-Uvarov method [2], tridiagonal J-matrix method [3], etc. Our motivation consists to the application of an alternative approach, based on Floquet theorem combined with the resonating average method [4] to determine the energy eigenvalues up to second order in the perturbation amplitudes, of cubic and quadric anharmonic oscillator. The analysis of this quantum anharmonic system, allowed us to obtain explicitly the expressions of vibration energy levels by introducing the Morse potential [5] and provide a useful description of diatomic molecules.

First, we give the theoretical context and the basic equations of our approach. Next, we apply it in case where the Morse potential function has developed to forth order. Therefore, we performed numerical calculations, using the established expressions as well as the parameters factors of the molecules species: LiH, NO and CO. The choice of these three molecules is for the following reasons: the molecules are different in terms of reduced mass and for the comparison with the results published in the literature. We determined the maximum quantum number energy level and we deuced the dissociation energies of the previous selected molecules.

The organisation of our paper is as follows: In section 2, we give the theoretical background and the general formulas. The section 3 contains the numerical results, comparisons with the literature and concluding comments. In section 4, we give the conclusions.

2. General Formulas

The Morse function [5] can be written as follows:

$$V(q) = D_e(1 - e^{-\rho q})^2 \tag{1}$$

Where $q = r - r_e$

Moreover, where ρ is a parameter that controls the width of the potential, D_e is the depth of the potential well, r is the distance between the atoms and r_e is the equilibrium bond distance.

Since the oscillations are of small amplitudes, we can do the Taylor development of $V(q)$ around q equal zero to forth order:

$$V(q) = D_e\rho^2 q^2 - D_e\rho^3 q^3 + \frac{7}{12} D_e\rho^4 q^4 \tag{2}$$

Let us consider the anharmonic oscillator which the Hamiltonian such as:

$$H(t) = \frac{p^2}{2m} + D_e\rho^2 q^2 - D_e\rho^3 q^3 + \frac{7}{12} D_e\rho^4 q^4 \tag{3}$$

Where P is the quantum momentum and m is reduced mass of the molecule.

The application of our approach to solve the Schrödinger equation according to the equation(3), leads to determine the second order energy levels as follows [6][7][8]:

$$^{(2a)}E_n = \hbar\omega_0\left(n+\frac{1}{2}\right) + \mu_2\frac{3}{4}\hbar\omega_0\left(2n^2+2n+1\right) - \mu_1^2\frac{15\hbar\omega_0}{4}\left(n^2+n+\frac{11}{30}\right) \tag{4}$$

$$-\mu_2^2\frac{\hbar\omega_0}{8}\left(34n^3+51n^2+59n+21\right)$$

$$\text{Where:} \quad \omega_0 = \rho\sqrt{\frac{2D_e}{m}} \quad , \mu_1 = -\frac{D_e\rho^3}{m\omega_0^2}\sqrt{\frac{\hbar}{m\omega_0}} \quad \text{and} \quad \mu_2 = \frac{7}{12}\frac{D_e\rho^4\hbar}{m^2\omega_0^3} \tag{5}$$

μ_1 and μ_2 are respectively the amplitudes of the cubic and quadric anharmonic perturbation.

We use the following identities:

$$\left(2n^2+2n+1\right) = 2\left(n+\frac{1}{2}\right)^2 + \frac{1}{2} \quad , \quad \left(n^2+n+\frac{11}{30}\right) = \left(n+\frac{1}{2}\right)^2 + \frac{7}{60}$$

$$\text{and} \quad \left(34n^3+51n^2+59n+21\right) = 34\left(n+\frac{1}{2}\right)^3 + \frac{67}{2}\left(n+\frac{1}{2}\right) \tag{6}$$

By replacing the identities of equation (6) into equation (4), we obtain the following equation:

$$^{(2a)}E_n = -\hbar\omega_0\left(\frac{7}{16}\mu_1^2 - \frac{3}{8}\mu_2\right) + \hbar\omega_0\left(1-\frac{67}{16}\mu_2^2\right)\left(n+\frac{1}{2}\right) - \hbar\omega_0\left(\frac{15}{4}\mu_1^2 - \frac{3}{2}\mu_2\right)\left(n+\frac{1}{2}\right)^2 \tag{7}$$

$$-\hbar\omega_0\frac{17}{4}\mu_2^2\left(n+\frac{1}{2}\right)^3$$

Let us note eplacing μ_1 and μ_2 by their expressions of equation (5) the term $\left(\frac{7}{16}\mu_1^2 - \frac{3}{8}\mu_2\right) = 0$.

Since the experimental data shows the terms in $\left(n+\frac{1}{2}\right)^3$ are negligible, we can write the analytical expression of the vibrational energy of diatomic molecules as follows:

$$^{(2a)}E_n = \hbar\omega_0\left(1-\frac{67}{16}\mu_2^2\right)\left(n+\frac{1}{2}\right) - \hbar\omega_0\left(\frac{15}{4}\mu_1^2 - \frac{3}{2}\mu_2\right)\left(n+\frac{1}{2}\right)^2 \tag{8}$$

Substituting ω_0, μ_1 and μ_2 in equation (8), we obtain the quasi-energies formula such as:

$$^{(2a)}E_n = \hbar\rho\sqrt{\frac{2D_e}{m}}\left[1-\frac{3283}{18432}\frac{\hbar^2\rho^2}{mD_e}\right]\left(n+\frac{1}{2}\right)-\frac{\hbar^2\rho^2}{2m}\left(n+\frac{1}{2}\right)^2 \tag{9}$$

Note that the exact molecular spectra are usually given by the following formula:

$$^{(ext)}E_n = \omega_e\left(n+\frac{1}{2}\right)-\omega_e\chi_e\left(n+\frac{1}{2}\right)^2 \tag{10}$$

Where ω_e and $\omega_e\chi_e$ are directly related to the Morse potential parameters. Thence from the equation (9), we can easily express the vibration frequency of the equilibrium ω_e and the first constant anharmonicity $\omega_e\chi_e$ as:

$$\omega_e = \hbar\rho\sqrt{\frac{2D_e}{m}}\left[1-\frac{3283}{18432}\frac{\hbar^2\rho^2}{mD_e}\right] \qquad \omega_e\chi_e = \frac{\hbar^2\rho^2}{2m} \tag{11}$$

3. Numerical results and comparisons

The Morse potential parameters and the spectroscopies constants of the selected molecules are given on table (1) [9][10].

Table 1. Parameters for the diatomic molecules LiH, NO and CO.

Molecules	D_e(eV)	m(uma)	ρ(Å^{-1})	ω_e(eV)	$\omega_e\chi_e$(eV)
LiH	2.515287	0.8801221	1.1280	0.174272	0.016504
NO	8.043729	7.46835	2.486737	0.236082	0.007391
CO	11.2256	6.8606719	2.2994	0.269015	0.006124

As illustrations of the equations (9) and (10), inserting the Morse parameters of the selected molecules and by using the Maple software, we determine the numerical values of the vibrational energy level and the comparison of our results with the so-called exact results in the following tables:

IOP Conf. Series: Materials Science and Engineering 1160 (2021) 012003 doi:10.1088/1757-899X/1160/1/012003

Table 2. Energy levels values and maximum quantum number of LiH.

n	E_n(eV)	
	Our results	Exact results
0	0.086333	0.086417
1	0.254469	0.254936
2	0.416560	0.417704
3	0.572608	0.574718
10	1.495720	1.512745
20	2.300737	2.363805
28	2.509605	2.630461
29	2.508517	2.637906
30	2.501385	2.639599
31	2.488210	2.635538
n_{max}	28	30

Table 3. Energy levels values and maximum quantum number of NO.

n	E_n(eV)	
	Our results	Exact results
0	0.117552	0.117605
1	0.350062	0.350197
2	0.579110	0.579300
3	0.804697	0.804913
10	2.286883	2.286480
30	5.587099	5.577219
60	7.941313	7.895800
67	8.042368	7.984842
68	8.042958	7.983602
69	8.040087	7.978873
n_{max}	68	67

Table 4. Energy levels values and maximum quantum number of the CO molecule.

n	E_n(eV)	
	Our results	Exact results
0	0.133361	0.134095
1	0.399081	0.399816
2	0.661578	0.662242
3	0.920854	0.921372
10	2.646138	2.643027
20	4.836043	4.822459
40	8.249337	8.192825
60	10.373942	10.245192
80	11.209858	10.979561
81	11.217826	10.981682
82	11.222572	10.980508
83	11.224096	10.976039
84	11.222399	10.968275
n_{max}	83	81

For testing our results, we present in **table 5**, the comparisons of numerical values of the maximum quantum number n_{max}, D_0 dissociation energy and the constants of anharmonicity obtained by using the quasi-energy expression of our approach, with the values calculated from the equation (10).
The maximum quantum number of vibration energy levels corresponds to the value of n beyond which the energy becomes maximum and the molecule will be dissociated, then n_{max} is the solution to the equation: $\frac{dE_n}{dn} = 0$, then we calculate the dissociation energy by using the expression :
$D_0 = E_{max} - E_0$ where E_0 is the zero point energy . The numerical values of the anharmonicity constants ω_e and $\omega_e\chi_e$ are calculated using the equation (11).
We use the established formulas (9), (10) and (11) to compute the maximum quantum number of vibrations n_{max} , the dissociation energy and spectroscopies constants relating to the parameters of the LiH, NO and CO molecules are given by [10].

Table 5. Numerical value of n_{max} , D_0 , ω_e and $\omega_e\chi_e$.

	Molecules	**LiH**	**NO**	**CO**
n_{max}	Our results	28	68	83
	Exact results	30	67	81
D_0 (eV)	Our results	2.425393	7.926653	11.091193
	Exact results	2.553182	7.867237	10.841943
ω_e (eV)	Our results	0.174201	0.235970	0.268924
	Exact results	0.174272	0.236082	0.269015
$\omega_e\chi_e$ (eV)	Our results	0.003025	0.001731	0.001610
	Exact results	0.002876	0.001745	0.001647

From the analysis of the results presented in the preceding table, we have shown that there is a concordance between the numerical values obtained using our approach, dismaying the maximum quantum number n_{max}, the dissociation energy D_0 and the constants of anharmonicity, and the numerical values established from the equation (10) so-called exact results.

IOP Conf. Series: Materials Science and Engineering 1160 (2021) 012003 doi:10.1088/1757-899X/1160/1/012003

4. Conclusion

We applied the analytical approach, to determine the quasi-energies up to the second order, solutions of the Schrodinger equation, in the case of the Morse potential, which is a suitable model to descript of the vibrations spectra of diatomic molecules.

From **tables 2, 3** and **4**, it can be seen that the calculation of energy levels values based on the formula (9) agree well, with those deduced from the Morse formula (10) when applying them to the previous diatomic molecule. Then we have applied the established analytical formulas to calculate some spectroscopic properties of diatomic systems in **table 5**. The comparisons of our numerical results with the literature are in good agreement and confirm the contribution of our approach.

We plan in future work to perform calculations by our approach, to determine the vibrational quasi-energies and the spectroscopies constants of anharmonicity where the Morse potential function will be developed to higher orders.

REFERENCES

[1] Flügge, S. (1994) Practical *Quantum Mechanics*, Springer-Verlag, New York, 68-84.

[2] M. Hamzavi[1], S. M. Ikhdair [2], Equivalence of the empirical shifted Deng-Fan oscillator potential for diatomic molecules, *Journal of Mathematical Chemistry* **51** (1) (2013) 227-238.

[3] Amlan K. Roy. Accurate ro-vibrational spectroscopy of diatomic molecules in a Morse oscillator potential, *Results in Physics* **3** (2013) 103–108.

[4] Janati-Idrissi, M., Fedoul, A., Achkar, Y., Chatwiti A. and Sayouri, S. (2011) Squeezing in Floquet States and Quasi-energies of Harmonic Oscillator Driven by a Strong Periodic Field, *Afr. J. Math. Phys.* ***10**, 21-30.*

[5] Morse, P. M. (1929) Diatomic molecules according to the wave mechanics. *II.Vibrational levels, Phys. Rev.***34**, 57-64.

[6] Janati-Idrissi, M., Fedoul, A, Sayouri, S. and Amila, I. (2020) Anharmonic Potentials Analysis through the Floquet Representation *Journal of Applied Mathematics and Physics*, **8**, 184-195.

[7] Wang, G., Guo, Q., Wu, L. and Yang, X. (2007) Giant Second-Order Optical Nonlinearities in Anharmonic Oscillator Potential Wells: Perturbation Theory Calculations. *Physica,* **39**, 75-84.

[8] Sun H. (2005) The Morse potential eigen energy by the analytical transfer matrix method. *Phys. Lett. A*, **338**, 309-314.

[9] Nasser, I., Abdelmonem, M.S., Bahlouli, H. and Alhaidari, A.D. (2007) The Rotating Morse Potential Model for Diatomic Molecules in the Tridiagonal J-Matrix Representation: I. Bound States. *Journal of Physics B: Atomic, Molecular and Optical Physics,* **40**, 4245-4257.

[10] *CRC Handbook of chemistry and physics*, Ed David R. Lide, 87th ed, Section 9, Spectroscopic constants of diatomic molecules p. 9-82.

IOP Conf. Series: Materials Science and Engineering 1160 (2021) 012004 doi:10.1088/1757-899X/1160/1/012004

Low temperature treatment and structural characterization of $Na_2M_2Fe(PO_4)_3$ (M= Mn or Ni) Alluaudite phases

Nour El Hoda Bouftila, Mohamed Naji*, Yahya Ababou, Abdelilah Rjeb, Salaheddine Sayouri

Université Sidi Mohamed Ben Abdellah, Faculté des Sciences Dhar El Mahraz, Laboratoire de Physique Appliquée Informatique et Statistique (LPAIS), Fez. B.P. 1796, Morocco.

Corresponding author: mohamed.naji3@usmba.ac.ma

Abstract. Materials for sodium intercalation were recently considered as attractive potential cathode materials for Na-ion Batteries. Amongst others, Alluaudite phases have shown interesting electrochemical properties. Thanks to their structure which consists of a three-dimensional framework of octahedra and tetrahedra sharing corners and/or edges, giving rise to two types of 1D tunnels where the Na+ ions can be intercalated. As first attempt to optimize the electrochemical performances of two alluadiates phases used as cathode materials, namely, Na2Mn2Fe(PO4)3 and Na2Ni2Fe(PO4)3,we have prepared these compounds by co-precipitation method with controlled nanometric size and low calcination temperature of 700°C and 750°C respectively. The structure of both samples obtained at different calcination temperatures were investigated using powder X-ray diffraction. The experimental results show the purity of Na2Mn2Fe(PO4)3 and Na2Ni2Fe(PO4)3 phases as well as the high crystallinity while maintaining low particle size and low calcination temperatures, namely less than 100 – 150°C difference.

1. Introduction

Energy and climate concerns have made the need for research towards electrical energy storage. In this context, sodium ion battery (SIBs) has attracted significant attention thanks to its huge availability, its low price and the similarity of both Li and Na insertion chemistries which makes the transition into using Na chemistry for ion battery systems feasible [1]. In spite of the lower energy density and voltage of Na-ion based technologies, they can be focused on applications where the weight and footprint requirement is less drastic, such as electrical grid storage that integrates discontinuous energy flow from renewable sources, optimizing the performance of clean energy sources[2]. In this regard much work has to be done in the field of Na-ion, cathodic and anodic materials must be optimized to find electrochemical couples that are suitable in terms of capacity and cycle life for NaIBs.

Thus, since the discovery of highly interesting properties of olivine phase LiFePO4 as a positive electrode material, several polyphosphate based compounds have been extensively studied as electro-positive materials in Na-ion batteries [3]. Among the large number of existing sodium phosphate phases, the alluaudite-type compounds have been extensively studied; thus, different research groups have recently reported diverse alluaudite compounds containing di and/or trivalent metals, with a monoclinic symmetry and C2/c space group[4][5],or an orthorhombic crystal structure with Imma space group with a more symmetric polyanionic framework.[6][7]

It is worth to mention that the alluaudite structure was first determined on natural minerals by Fisher[8], who showed that alluaudite compounds crystallize in the monoclinic C2/c space group. Whereas the general formula X(2)X(1)M(1)M(2)2(PO4)3, was proposed by Moore [9]in which X and M cations are written according to decreasing size. where its structure consists of a three-dimensional framework made up of infinite chains formed by a succession of octahedral M (2) pairs connected by highly deformed M (1) octahedra sharing edges, and the equivalent chains are connected by phosphate tetrahedra to form sheets oriented perpendicular to [010] than equivalent sheets are held together as well by PO4 tetrahedra to form a three-dimensional architecture with two sets of tunnels in the c direction: tunnel 1 (1/2, 0, z) and tunnel 2 (0, 0, z) where X1 and X2 are cations residing in different sites of these tunnels.

The electrochemical performances are strongly affected by the synthesis condition for the sake of difference in homogeneity and for their dependency on the size and the morphology of the particles [10][11]. Therefore, the control of homogeneity and integrity of the host structure are most important factors to obtain better electrochemical properties such as high reversible capacity and stable cycling performance. In this paper, we report on the synthesis and characterization of $Na_2Mn_2Fe(PO4)_3$ and $Na_2Ni_2Fe(PO_4)_3$prepared by co-precipitation method, and calcinated at different high temperatures. As compared to literature, the co-precipitation method allowed us to reduce the calcination temperature by up to 150°C while maintaining very low nanometric particle size and good crystallinity. [12]

2. Experimental Section

2.1 Synthesis

Both samples of $Na_2Mn_2Fe(PO_4)_3$ and $Na_2Ni_2Fe(PO_4)_3$, were prepared by the co- precipitation method from mixing an appropriate amounts of Na_2CO_3 (99.5%, Sigma-Aldrich), $Fe(NO_3)_3.9H_2O$ (99%, Sigma-Aldrich) and $Mn(CH_3CO_2)_2 \cdot 4H_2O$ (99%, Sigma-Aldrich) or $Ni(CH_3CO_2)2 \cdot 4H_2O$ (99%, Sigma-Aldrich) in distilled water under vigorous magnetic stirring, then the required amount of $NH_4H_2PO_4$ (99%, Panreac - AppliChem) dissolved in distilled water was slowly added to the mixture and stirred for 12h. The excess water was then evaporated by heating at 70°C and the residual powders were ground in an agate mortar then placed in a muffle furnace and heated at 200°C for 24h with a heat rate of 5°C per min, then at 400°C for 24h to get rid of organic chemical compounds and finally calcinated at 600°C, 700°C, 750°C, 800°C and 850°C for 48h. At the end of the synthesis, the powder obtained by final quenching of the product is pale yellow for nickel and brown for manganese compound.

2.2 X-ray diffraction

The X-ray powder diffraction patterns were recorded using a Panalytical diffractometer operating with CuKα radiations at a wavelength of $K\alpha 1 = 1.54060$ Å and $K\alpha 2 = 1.54443$ Å, and the data were collected over the 2θ angle range of $10° \leq 2\theta \leq 80°$ with a step size of 0.017°.

3. Characterization analysis

3.1 Phase analysis of materials

X-ray powder diffraction was used in this work to identify crystalline phases and to analyze their structures, as shown in Figures 1 and 2.The analysis of X-ray diffraction spectra of the compound $Na_2Mn_2Fe(PO_4)_3$ carried out at different temperatures showed that the compound is pure can be indexed in the monoclinic space group C2/c. Further, as it can be seen from Figure 1, that at higher calcination temperatures, the Bragg peaks become sharper and clearer, indicating that the crystallinity is increasing.

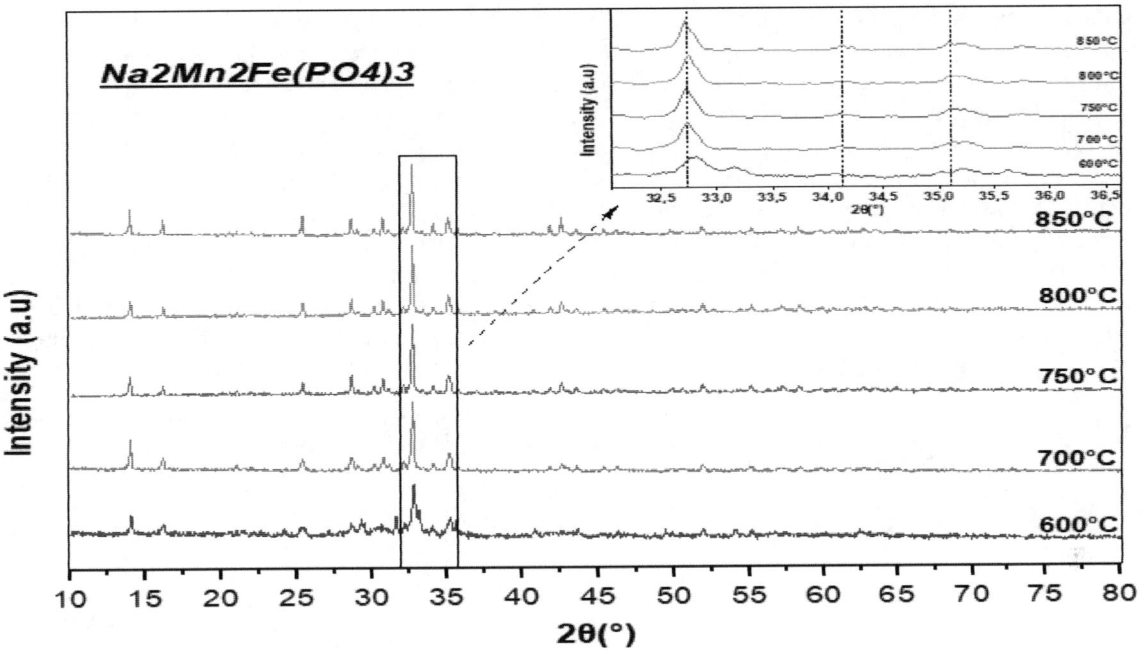

Figure 1. X-ray powder diffraction patterns of $Na_2Mn_2Fe(PO_4)_3$ calcinated at different temperatures. The inset corresponds to a zoom of the 2θ = 32 to 36.6° area.

The particle size of the crystallites was calculated using the Scherrer equation according to the following relation: $D = \frac{K\lambda}{\beta \cos(\theta)}$; Where (D is the size of crystals; K is a dimensionless shape factor, with typical value of about 0.9; λ is the X-ray wavelength; β is the line broadening at half the maximum intensity (FWHM); θ is the Bragg angle).
Table 1 and 2 report the measured Full width at half maximum (FWHM) of two intense peaks from the X-ray patterns, as well as the calculated average particles size.

Table 1. Full width at half maximum (FWHM) of the most intense peaks (-240) and (-132)

$Na_2Mn_2Fe(PO_4)_3$		FWHM				
hkl	**2θ**	600°C	700°C	750°C	800°C	850°C
(-240)	32.74664	0.57017	0.20785	0.18659	0.17629	0.1666
(-132)	35.18682	0.65292	0.28529	0.27127	0.24849	0.25319
D_{avg} (nm)		136.6021	345.6771	375.9587	403.0382	413.5976

Thus we can notice from table 1 and figure 3 that, with the increase of the calcination temperature, the width of diffraction peaks is lowered, while the average size increase in the range between 136 and 413 nm. This clearly indicates that the high temperature calcination induce a rise in the agglomeration of the powder and an increase in the average particle size. Hence we opted for a calcination at 700°C due to the good crystallinity of the powder, and a relatively small particle size. Whereas its analogue was synthesized by a solid-state reaction carried out in air and progressively heated to 850°C, in order to obtain high-purity alluaudite [13]. In that the possibility of improving their performance through particle size reduction is being investigated, owing to important role that plays this latter in the electrochemical performances.[10]

IOP Conf. Series: Materials Science and Engineering 1160 (2021) 012004 doi:10.1088/1757-899X/1160/1/012004

Similarly, the investigation of the diffractogram of the $Na_2Ni_2Fe(PO_4)_3$ phase indicates that this compound crystallizes in the orthorhombic structure with the space group Imma, which is in concordance with previous work [14].As for the manganese counterpart, the choice of the optimum temperature, the influence of crystallinity and of agglomeration factors have been taken into account exhaustively, and as it can be seen from Figure 2, the characteristic diffraction peaks of the different spectra corresponding to calcination temperatures ranging from 600°C to 850°C, exhibit a slight difference namely, in crystallinity. Therefore, $Na_2Ni_2Fe(PO_4)_3$ powder calcined at 750°C was chosen, where the low intensity peaks appear more clearly than in the spectrum of 700°C with high cristalinitty, and rather lower than the high temperature of 850°C required for the synthesis by solid state reaction [14]. Ultimately, a decrease of full width at half maximum (FWHM) of the peaks as the temperature rises was noticed (Table 2)therefore we estimate through the widths measured and by using the Scherrer formula, an average size ranging between 177 and 463 nm (Figure 3), confirming that the higher temperature certainly lead to more severe powder agglomeration.[15]

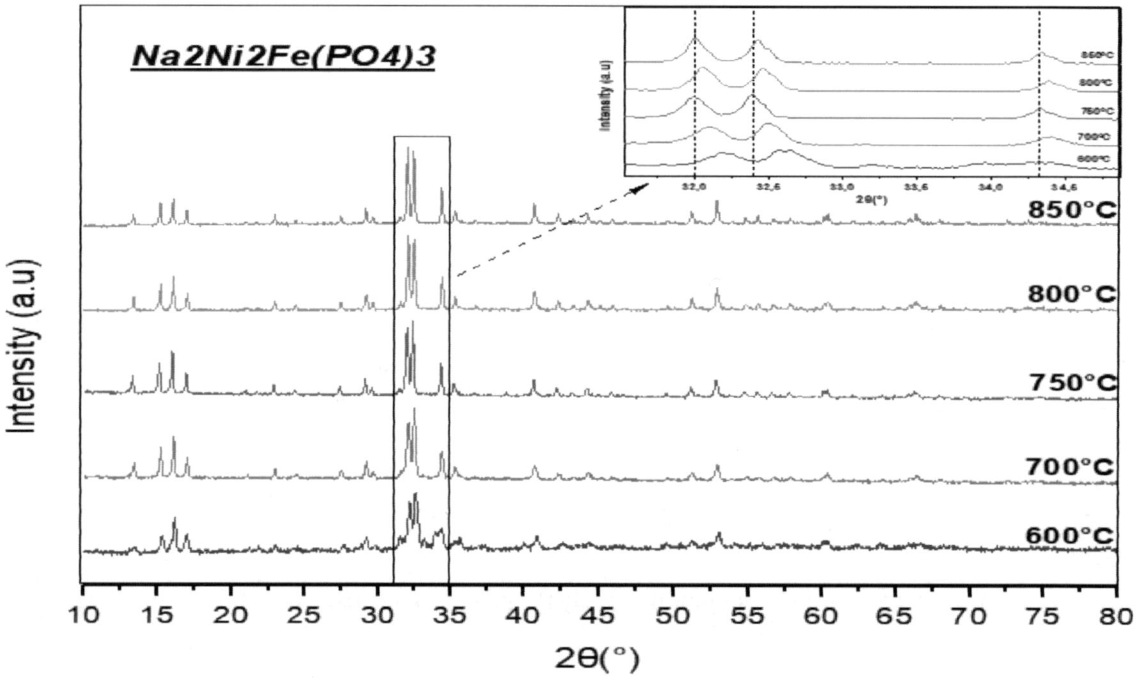

Figure 2. X-ray powder diffraction patterns of $Na_2Ni_2Fe(PO_4)_3$ calcinated at different temperatures. The inset corresponds to a zoom of the 2θ = 31.5 to 34.8° area.

Table 2. Full width at half maximum (FWHM) of (400) and (251) peaks

$Na_2Ni_2Fe(PO_4)_3$		FWHM				
hkl	**2θ**	600°C	700°C	750°C	800°C	850°C
-400	34.40061	0.65691	0.22961	0.17766	0.18418	0.1597
-251	40.75677	0.34147	0.2954	0.22598	0.22361	0.18513
D_{avg} **(nm)**		177.0117	307.8440	399.8518	393.7924	463.861

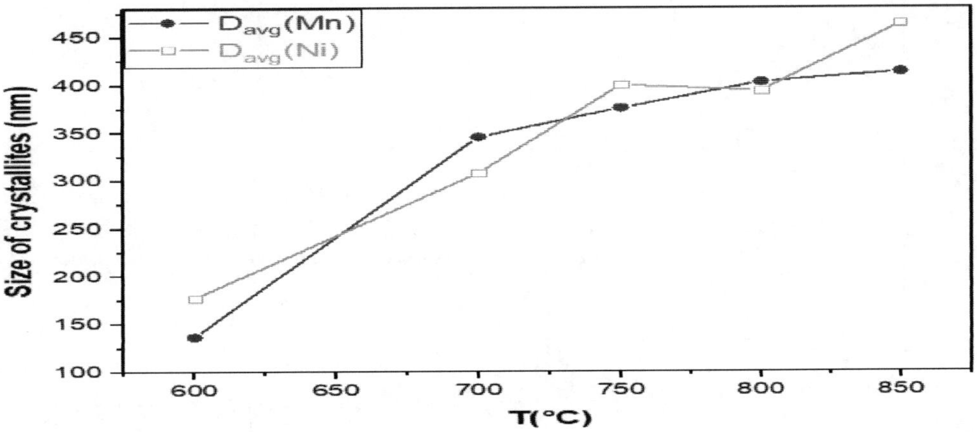

Figure 3. The average size crystals of $Na_2Mn_2Fe(PO_4)_3$ and $Na_2Ni_2Fe(PO_4)_3$ samples calcinated at different temperatures.

3.2 Structure refinement

The unit-cell parameters and the atomic positions reported for $Na_2Mn_2Fe(PO_4)_3$, and for $Na_2Ni_2Fe(PO_4)_3$ have served as the starting parameters for the Rietveld refinement of the synthesised compounds which were performed with Fullprof program [13] [14]. The Rietveld refinement plot of the two synthesised phases is shown in Figure 4 and the refined cell parameters are listed in Table 3.

Table 3. Crystallographic and structure refinements data

Rietveld Refinement		
Structural formula	$Na_2Mn_2Fe(PO_4)_3$	$Na_2Ni_2Fe(PO_4)_3$
Molecular weight	496.61	504.11
Crystal system	Monoclinic	Orthorhombic
Space group	C2/c (No 15)	Imma (No 74)
Cell dimension (Å)	a=12.0311(13)	a=10.4063(17)
	b=12.6122(14)	b=13.195(2)
	c=6.4972(6)	c=6.4882(11)
	β=114.492(7)	β = α = γ = 90°
Cell volume (Å³)	897.16(16)	890.9(3)
Z	2	4
goodness of fit χ2	0.7745	1.04
Agreeing factors	Rp = 49.2	Rp = 36.2
	Rwp = 33.1	Rwp = 22.5
	Rexp = 37.6	Rexp = 22.02
No. of parameters	56	37
Profile function	Pseudo-Voigt	
Background	6-coefficients polynominal function	

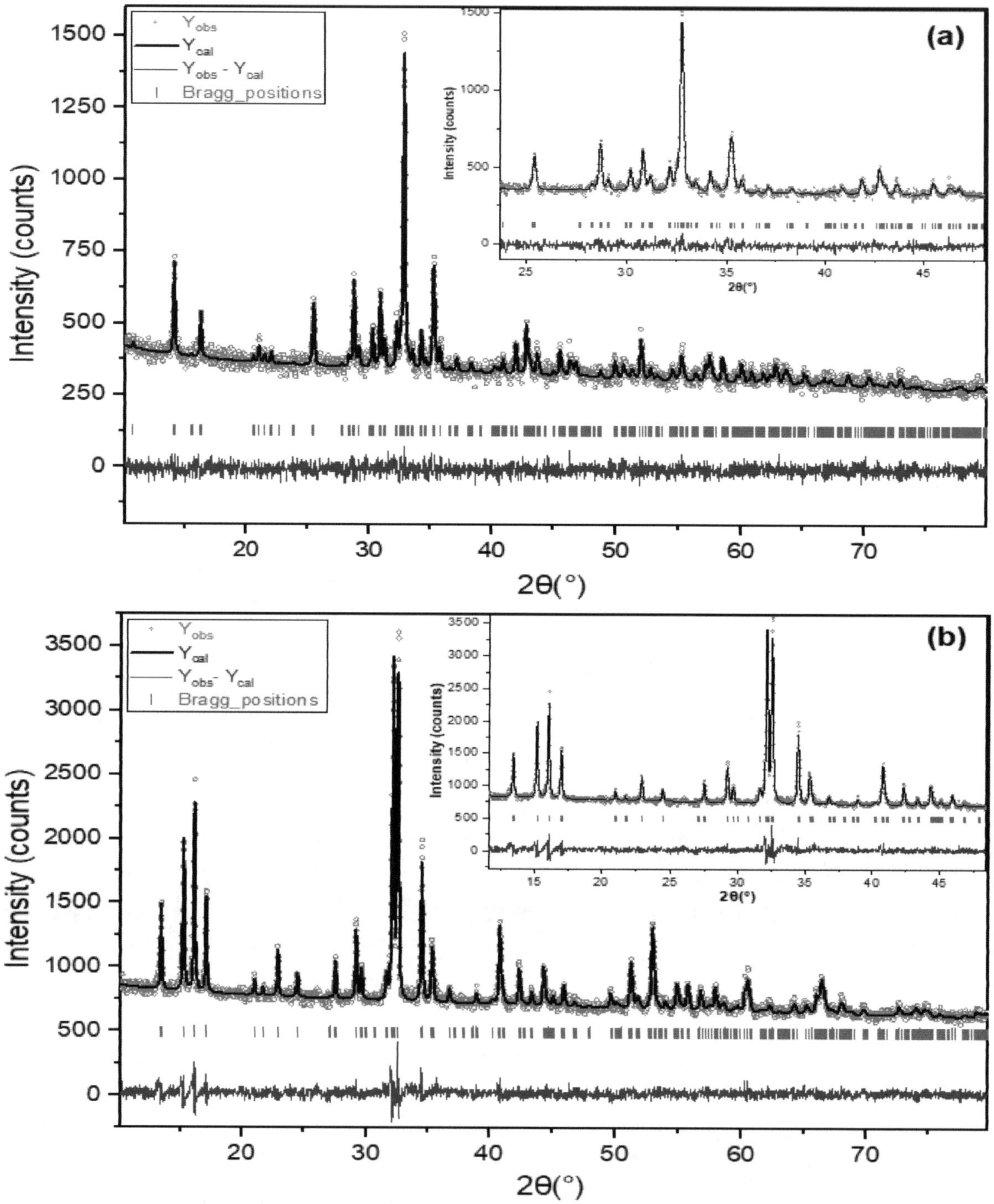

Figure 4. The observed, calculated, and difference X-ray powder diffraction patterns of $Na_2Mn_2Fe(PO_4)_3$ (a) and $Na_2Ni_2Fe(PO_4)_3$ (b) obtained from a Rietveld refinement.

One can notice from the Rietveld refinement of $Na_2Mn_2Fe(PO_4)_3$ and $Na_2Ni_2Fe(PO_4)_3$compounds (Figure 4) that the refinement converges towards a good agreement between the experimental and calculated diffractograms, which confirms the adequacy of the structural model.

From the Rietveld analysis the bond distance analysis for Na-O, Fe-O, Ni-O and Mn-O gives rather satisfactory results, in agreement with the theoretical values as calculated from Shannon ionic radii table [16] (see Tables 4 and 5).

Table 4. Selected interatomic distances (Å) determined for $Na_2Mn_2Fe(PO_4)_3$ from the Rietveld Refinement of the XRD data.

Na(2)' - O(1)	3.27(5)	Na(1) - O(2) × 2	2.28(4)	Mn(2)/Fe(2) - O(1)	2.40(4)
Na(2)' - O(1)	1.36(5)	Na(1) - O(2) × 2	3.01(5)	Mn(2)/Fe(2) - O(2)	1.51(4)
Na(2)' - O(2)	2.42(4)	Na(1) - O(4) × 2	2.38(2)	Mn(2)/Fe(2) - O(3)	3.08(5)
Na(2)' - O(3)	2.19(5)	Na(1) - O(4) × 2	2.45(4)	Mn(2)/Fe(2) - O(4)	2.64(4)
Na(2)' - O(3)	2.59(4)	< Na(1) - O>	2.53	Mn(2)/Fe(2) - O(5)	1.30(3)
Na(2)' - O(4)	2.67(3)			Mn(2)/Fe(2) - O(6)	1.75(5)
Na(2)' - O(4)	2.56(4)	Mn(1) - O(2)	2.54(5)	< Mn(2)/Fe(2) - O>	2.11
Na(2)' - O(5)	3.27(3)	Mn(1) - O(3) × 2	1.66(3)		
<Na(2)’ - O>	2.54	Mn(1) - O(4)	2.54(3)	P(2) - O(3)	1.51(4)
		Mn(1) - O(4)	2.62(2)	P(2) - O(4)	1.58(4)
P(1) - O(1) × 2	1.52(4)	Mn(1) - O(5)	1.87(4)	P(2) - O(5)	1.54(5)
P(1) - O(2) × 2	1.63(5)	<Mn(1) - O>	2.24	P(2) - O(6)	1.57(4)
<P(1) - O>	1.57			<P(2) - O>	1.55

Table 5. Selected interatomic distances (Å) determined for $Na_2Ni_2Fe(PO_4)_3$ from the Rietveld Refinement of the XRD data.

Na(1) - O(1) × 4	2.664(9)	Ni(1)/Fe(1) - O(1) × 2	2.106(3)
Na(1) - O(4) × 2	2.747(7)	Ni(1)/Fe(1) - O(2) × 2	2.073(5)
Na(1) - O(3) × 2	2.587(6)	Ni(1)/Fe(1) - O(4) × 2	2.047(9)
<Na(1) - O>	2.666	<Ni(1)/Fe(1) - O>	2.075
Na(2) - O(1) × 4	2.882(5)	Ni(2)/Fe(2) - O(2) × 4	2.006(4)
Na(2) - O(2) × 4	2.374(5)	Ni(2)/Fe(2) - O(3) × 2	2.025(6)
<Na(2) - O>	2.628	<Ni(2)/Fe(2) - O>	2.015
P(1) - O(1) × 2	1.522(4)	P(2) - O(3) × 2	1.524(4)
P(1) - O(1) × 2	1.552(9)	P(2) - O(4) × 2	1.558(3)
<P(1) - O>	1.537	<P(2) - O>	1.541

3.3 Crystal structure

The following polyhedral representation (Figure 5) illustrates a projection view of the crystal structure of the title phase along (a) and (b) directions. Where we notice very well that the structure is formed by a linear sequence of MnO6 octahedron with Fe06 octahedra, forming a tri-dimensionnel structure mainly by phosphate tetrahedra and thus giving rise to cavities in the c direction where the sodium ions are inserted.

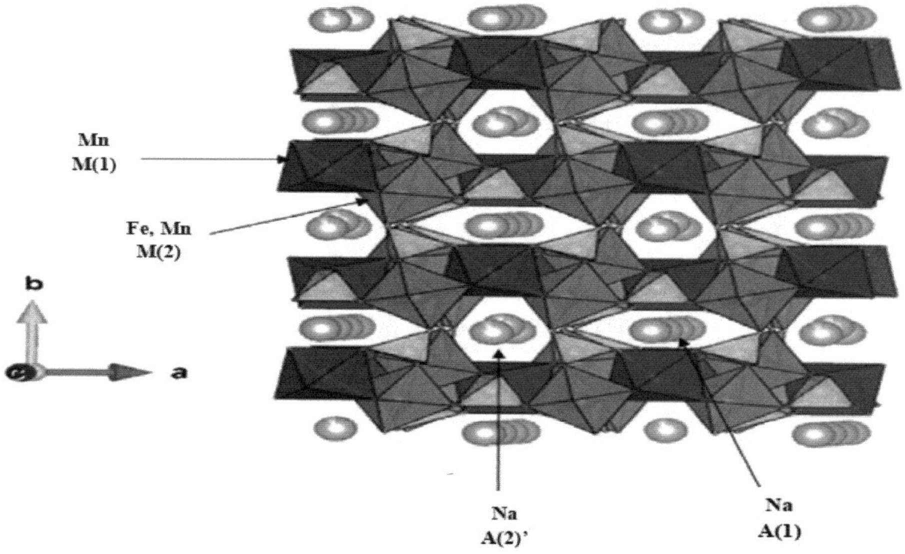

Figure 5. Projections of the monoclinic structure of $Na_2Mn_2Fe(PO_4)_3$ along the c axis.

The orthorohmbic structure (Figure 6) of the nickel compound can be easily described by considering the association of two anionic building entities, the first entity composed of infinite chains parallel to the $b^{\rightarrow}$ axis, obtained by the iron octahedra FeO6 and the tetrahedra of phosphate. While the second entity formed by the association of two Nickel octahedral dimers sharing edges with phosphate tetrahedra. Consequently, these entities share the corners of oxygen with the adjacent units to form an infinite layer, associated in its turn to form a three-dimensional anionic frame with the creation of two channels along the directions $a^{\rightarrow}$ and $b^{\rightarrow}$, occupied by the Na^+ cations as shown in the crystal structure.

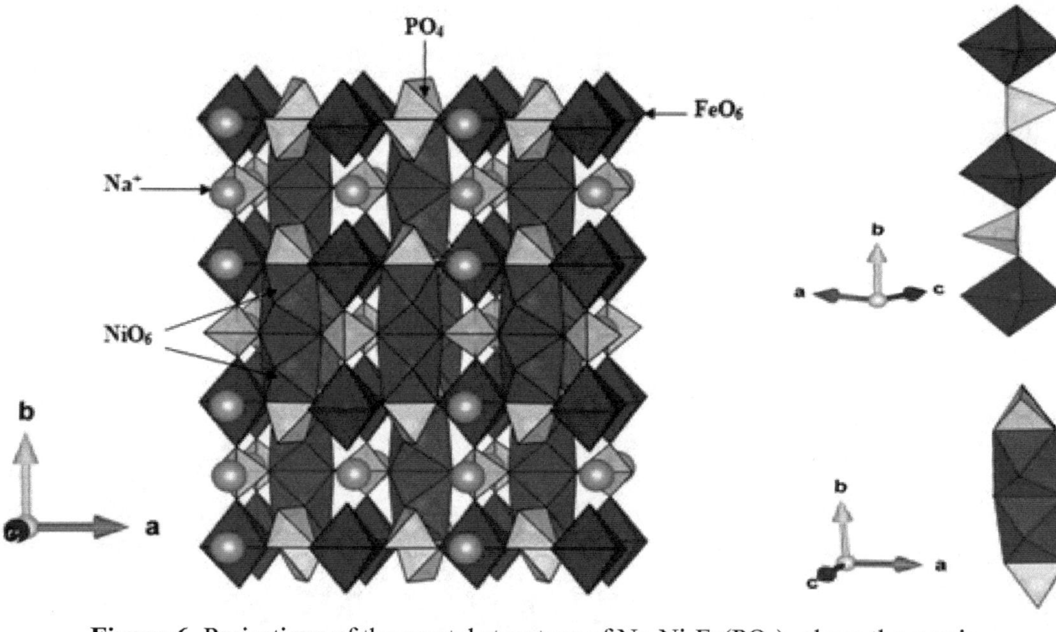

Figure 6. Projections of the crystal structure of $Na_2Ni_2Fe(PO_4)_3$ along the c axis.

Conclusion

Two Alluaudite phases $Na_2M_2Fe(PO_4)_3$ with (M=Mn, Ni) were prepared by co-precipitation method, and the Rietveld refinement of their XRD data indicate that this phases crystallised in the monoclinic system for manganese and orthorhombic for nickel compounds.The powders at different calcination temperatures were characterized by XRD analysis and then compared by considering the crystallinity and the effect of high temperature on the size of crystallites, as a smaller grain size is favorable to the sodium intercalation process in the primary particles by reducing the length of the ionic diffusion path. $Na_2Mn_2Fe(PO_4)_3$ calcinated at 700°C and $Na_2Ni_2Fe(PO_4)_3$ calcinated at 750°C were selected as optimum temperatures to explore their effectiveness as cathode materials for Na-ion battery.

References

[1] V. Palomares, P. Serras, I. Villaluenga, K. B. Hueso, J. Carretero-González, et T. Rojo, « Na-ion batteries, recent advances and present challenges to become low cost energy storage systems », *Energy Environ. Sci.*, 2012.
[2] M. Sawicki et L. Shaw, « Advances and challenges of sodium ion batteries as post lithium ion batteries REVIEW », *RSC Adv.*, 2015.
[3] J. Kang *et al.*, « High rate performance of a Na3V2(PO4)3/C cathode prepared by pyro-synthesis for sodium-ion batteries », *J. Mater. Chem.*, 2012.
[4] F. Hatert, D. Antenucci, A.-M. Fransolet, et M. Liégeois-Duyckaerts, « The Crystal Chemistry of Lithium in the Alluaudite Structure: A Study of the (Na1−xLix)CdIn2(PO4)3 Solid Solution (x=0 to 1) », *Journal of Solid State Chemistry*, 2002.
[5] R. Essehli et al, « Alluaudite $Na_2Co_2Fe(PO_4)_3$ as an electroactive material for sodium ion batteries », *Dalton Trans.*, 2015.
[6] K. Souiwa, M. Chennabasappa, R. Decourt, M. Ben Amara, M. Hidouri, et O. Toulemonde, « Novel Mixed Cobalt/Chromium Phosphate NaCoCr2(PO4)3 Showing Spin-Flop Transition », *Inorg Chem*, 2015.
[7] D. Harbaoui, M. Sanad, C. Rossignol, E-K. Hlil, N. Amdouni and S. Obbade, « Synthesis and Structural, Electrical, and Magnetic Properties of New Iron–Aluminum Alluaudite Phases β-Na2Ni2M(PO4)3 (M = Fe and Al) | Inorganic Chemistry », 2017.
[8] D. J. Fisher, « Alluaudites and Varulites », *American Mineralogist*, 1957.
[9] Moore et al, « Crystal Chemistry of The Alluaudite Structure Type: Contribution to the Para-Genesisof Pegmatite Phosphategiant Crystals », 1971.
[10] C. Benoit, « Étude des propriétés électrochimiques de nouveaux matériaux nanostructurés à base de fer préparés par chimie douce et utilisables comme électrodes positives d'accumulateurs au lithium ».
[11] D. Cheng Li, T. Muta, L.-Q. Zhang, M. Yoshio, et H. Noguchi, « Effect of synthesis method on the electrochemical performance of LiNi1/3Mn1/3Co1/3O2 », *Journal of Power Sources*, 2004.
[12] A. V. Rane, K. Kanny, V. K. Abitha, et S. Thomas, « Methods for Synthesis of Nanoparticles and Fabrication of Nanocomposites », in *Synthesis of Inorganic Nanomaterials*, 2018.
[13] F. Hatert *et al.*, « A structural, magnetic, and Mossbauer spectral study of several Na-Mn-Fe-bearing alluaudites », *Phys Chem Minerals*, 2004.
[14] R. Essehli, « α-Na2Ni2Fe(PO4)3: a Dual Positive/Negative Electrode Material for Sodium Ion Batteries », *Dalton Trans.*, 2015.
[15] A. Singhal, G. Skandan, A. Wang, N. Glumac, B. H. Kear, et R. D. Hunt, « On nanoparticle aggregation during vapor phase synthesis », *Nanostructured Materials*, 1999.
[16] R. D. Shannon, « Revised effective ionic radii and systematic studies of interatomic distances in halides and chalcogenides », *Acta Cryst A*, 1976.

Sol-gel synthesis and structural study of a lithium titanate phase $Li_{3x}La_{2/3-x}\square_{1/3-2x}TiO_3$ as solid electrolyte

Abdelhak Chouiekh, Mohamed Naji*, Abdelilah Rjeb, Yahya Ababou and Salaheddine Sayouri

Université sidi Mohamed Ben Abdellah (USMBA)
Laboratoire de Physique Appliquée, Informatique et Statistique (LPAIS), FSDM, B.P.1796, Fès-Atlas, Maroc

*Corresponding author: mohamed.naji3@usmba.ac.ma

Abstract. Perovskite lithium lanthanum titanate (LLTO) was prepared as an inorganic solid electrolyte material for lithium ion batteries applications using sol-gel method. Three different compositions of : (a) $3x$=0.3 ($Li_{0.3}La_{0.57}TiO_3$), (b) $3x$=0.5 ($Li_{0.5}La_{0.5}TiO_3$) were prepared and calcined at 1000°C for 6h, and (c) $3x$=0.1 ($Li_{0.1}La_{0.63}TiO_3$) prepared and calcined at 1100°C for 6h. The effects of lithium content and calcination temperature on the structure of the perovskite powders were studied. The synthesized powders were characterized by X-ray diffraction analysis. Results indicate that the $Li_{0.3}La_{0.57}TiO_3$ was crystallized in tetragonal structure of space group P4/mmm. On the other hand, Rietveld refinement analysis for the LLTO samples confirmed the formation of the $Li_{0.3}La_{0.57}TiO_3$ crystalline phase in the tetragonal P4/mmm space group, the $Li_{0.5}La_{0.5}TiO_3$ phase in the tetragonal P4/mmm space group, and $Li_{0.1}La_{0.63}TiO_3$ phase in orthorhombic Pmmm space group.

Keywords
Perovskite; Lithium lanthanum titanate; All-solid-state battery; Sol-gel; LLTO; Refinement.

1. Introduction

All-solid-state batteries lithium are of considerable interest for future technological applications, such as portable electronics, micro-devices and even large-scale applications [1].

Lithium-ion batteries currently represent the main source of power in electronic devices and are widely regarded as one of the most remarkable technologies in transportation electrification. However, lithium-ion batteries still have some limitations, particularly in terms of security. These security problems are often associated with the use of flammable organic electrolytes [2-3].

One of the solutions to these problems is precisely the introduction of solid electrolytes. The advantages of using solid electrolytes instead of organic liquid electrolytes, besides security, are their good thermal stability, higher energy capacity due to the possibility of more compact structures without the need for separators as in the case of conventional batteries, and their wide range of operating temperatures [4].

One of the conditions for the application of the solid electrolyte is that it must have a reasonably higher ionic conductivity. The one potential solid electrolyte is ceramic perovskite ABO_3, such as lithium lanthanum titanate, of the formula $\mathbf{Li_{3x}La_{2/3-x}\square_{1/3-2x}TiO_3}$ called LLTO, (□ represents vacancy) [5]. The highest bulk ionic conductivity reported for LLTO is about 10^{-3} $S.cm^{-1}$ at room temperature when $3x = 0.3$ [6], but, the total ionic conductivity of LLTO is low (10^{-5} $S.cm^{-1}$) due to the low ionic conductivity at the grain boundary [7-8]. Different publications suggest that sintering temperatures, composition (Li to La ratio) and preparation method affect the structure and morphology of LLTO, modifying the resulting conductivity [4-9]. That way, although many reports have been made on the properties of LLTO, different preparation steps and sintering conditions may result in different or improved properties.

LLTO is generally synthesized by the sol-gel method or by solid-state reaction, the latter being easier and more economical. In general, the sintering temperature is above 1250°C, which leads to the evaporation of lithium [10]. In this study, the LLTO samples $\mathbf{Li_{3x}La_{2/3-x}\square_{1/3-2x}TiO_3}$ ($3x$=0.1; $3x$=0.3 and $3x$=0.5) were prepared by the sol-gel method, and characterized by XRD diffraction and their structures were refined using the Rietveld method.

2. Experimental

Figure 1. shows the diagram of the procedure for the synthesis of lithium lanthanum titanate (LLTO) powders. The sol-gel method was used to prepare three different compositions of LLTO for $3x$=0.1; $3x$=0.3 and $3x$=0.5 .

The starting materials were lanthanum (III) acetate hydrate $La(CH_3COO)_3.xH_2O$ (Sigma-Aldrich, 99% purity), titanium (IV) isopropoxyde $Ti[OCH(CH_3)_2]_4$ (Sigma-Aldrich, 97% purity) and lithium acetate $Li(CH_3COO).2H_2O$ (Sigma-Aldrich, 99% purity) as precursors, lactic acid $CH_3CH(OH)COOH$ (Sigma-Aldrich, 90% purity) as peptizing agent, and distilled water as solvent.

Lanthanum acetate, lithium acetate and titanium sol are mixed in stoichiometric proportions, according to the chemical formulation $\mathbf{Li_{3x}La_{2/3-x}\square_{1/3-2x}TiO_3}$ ($3x$=0.1; $3x$=0.3 and $3x$=0.5) under stirring for 2 hours, to ensure the homogeneity of the final solution. The destabilization of this solution is ensured by evaporation of the solvent in an oven at a temperature of (60°C ~ 70°C) for 96 hours. The xerogel obtained is ground in an agate mortar to break up the agglomerates of the powder and increase its reactivity. The mixed powders were calcined at 1000°C (6h) for ($3x$=0.3 and $3x$=0.5) and at 1100°C (6h) for ($3x$=0.1) was calcined.

The obtained samples were analyzed by X-ray diffraction using a computer-controlled XPERT-PRO diffractometer with a copper anode K-Alpha1 ($\lambda1$=1.54060Å) and K-Alpha2 ($\lambda2$=1.54443Å). A scan was adopted with a 0.0170° step. The chosen measuring range is from 10° to 90°.

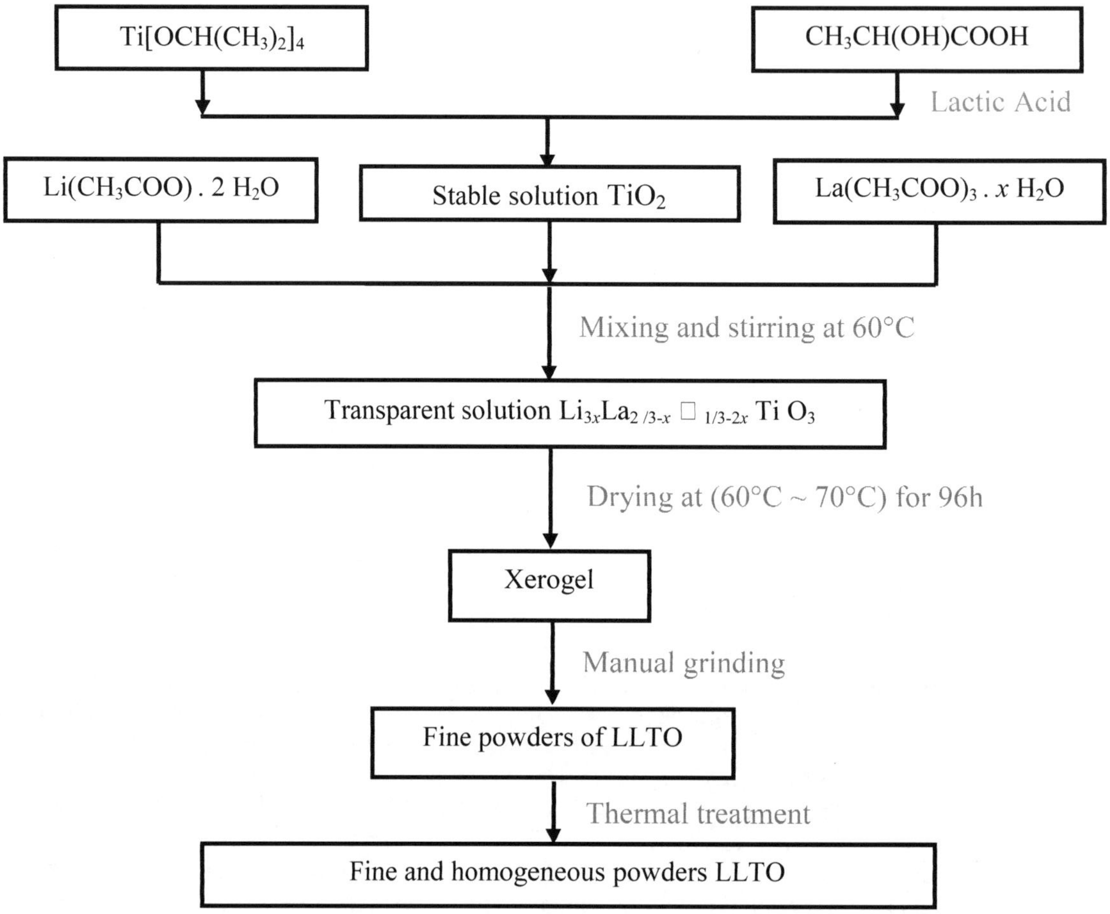

Figure 1. Schematic illustration of the LLTO synthesis by the sol-gel method

3. Results and discussion

3.1. Lithium lanthanum titanate for (3x=0.3) $Li_{0.3}La_{0.57}TiO_3$

The X-ray diffraction patterns of $Li_{0.3}La_{0.57}TiO_3$ obtained by sol-gel after heat treatments at 900°C, 950°C and 1000°C for 6 hours are shown in figure 2.

The formation of the LLTO phase starts at 900°C with impurities. The main impurity is the lanthanum titanium oxide $La_2Ti_2O_7$ which was also found by Bohnke et al [11] during the preparation of $Li_{0.33}La_{0.56}TiO_3$ by the sol-gel route and the lithium titanate Li_2TiO_3 which was found by Romero et al [12] during the preparation of $Li_{0.3}La_{0.57}TiO_3$. When the temperature of the treatment increases, the proportion of impurities decreases, and the calcination temperature at 1000°C promotes the removal of these impurities. $La_2Ti_2O_7$ appears at 700°C as indicated by Kitakoa et al [13].

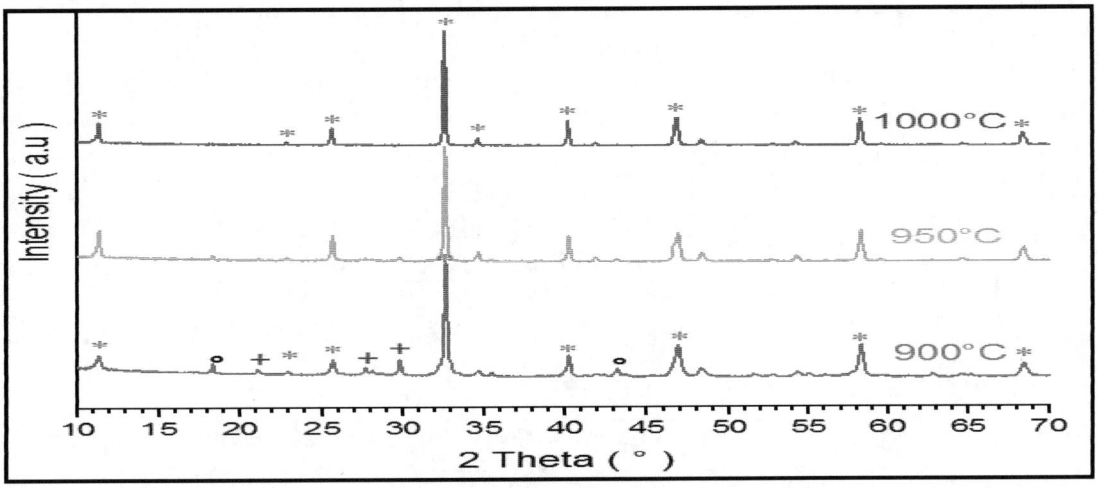

Figure 2. X-ray diffraction of $Li_{0.3}La_{0.57}TiO_3$ obtained by sol–gel method; (*) peaks of LLTO(3x=0.3); (**o**) peaks of Li_2TiO_3 and (+) peaks of $La_2Ti_2O_7$

The structure of $Li_{0.3}La_{0.57}TiO_3$ heat treated at 1000°C for 6 hours was further refined by means of Rietveld method (Figure 3). In a first step, a refinement was performed using the structural model proposed in the literature by Fourquet et al. [14], which allowed us to verify that the compound is pure LLTO (3x=0.3), and that it crystallizes well in a tetragonal cell (P4/mmm; a=b=3.8704Å and c=7.7621Å and α=β=γ=90°). Reliability factors obtained after refinement (R_p=26.80%; R_{wp}=20.30%; R_{exp}=15.17% and χ^2=1.79). In table 1, the crystallographic characteristics of $Li_{0.3}La_{0.57}TiO_3$ are presented, and in table 2, the interatomic distances are reported for refined LLTO (3x=0.3), these results are in very good agreement with the literature [15].

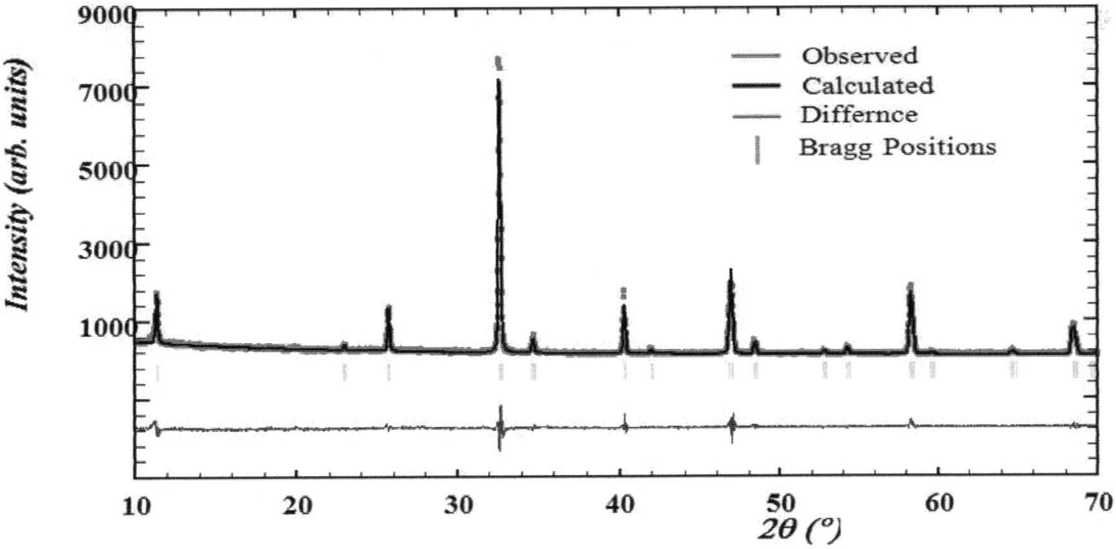

Figure 3. Shows the result of the refinement performed with the Fullprof program with the comparison of the calculated and observed diagrams for LLTO (3x=0.3)

Table 1. Wyckoff Sites, Atomic coordinates, for the *P4/mmm* Structure of $Li_{0.3}La_{0.57}TiO_3$

Atom	Site	Atomic coordinates			Occ
		x	y	z	
La1	1a	0.00000	0.00000	0.00000	0.05374
La2	1b	0.00000	0.00000	0.50000	0.02168
Ti	2h	0.50000	0.50000	0.26058	0.12500
O1	1c	0.50000	0.50000	0.00000	0.06250
O2	1d	0.50000	0.50000	0.50000	0.06250
O3	4i	0.00000	0.50000	0.24240	0.25000

Table 2. Interatomic Distances in Å of $Li_{0.3}La_{0.57}TiO_3$

<dTi-O1>	<dTi-O2>	4 × <dTi-O3>	4 × <dLa1-O1>
2.0226	1.8584	1.9403	2.7368

8 × <dLa1-O3>	4 × <dLa2-O2>	8 × <dLa2-O3>
2.6991	2.7368	2.7826

3.2. Lithium lanthanum titanate for (3x=0.5) $Li_{0.5}La_{0.5}TiO_3$

For LLTO samples (3*x*=0.5), the X-ray diffractogram registered on the heat-treated compound at 1000°C is shown in figure 4. The LLTO compound has indeed been formed, with the presence of impurity containing lithium titanate Li_2TiO_3.

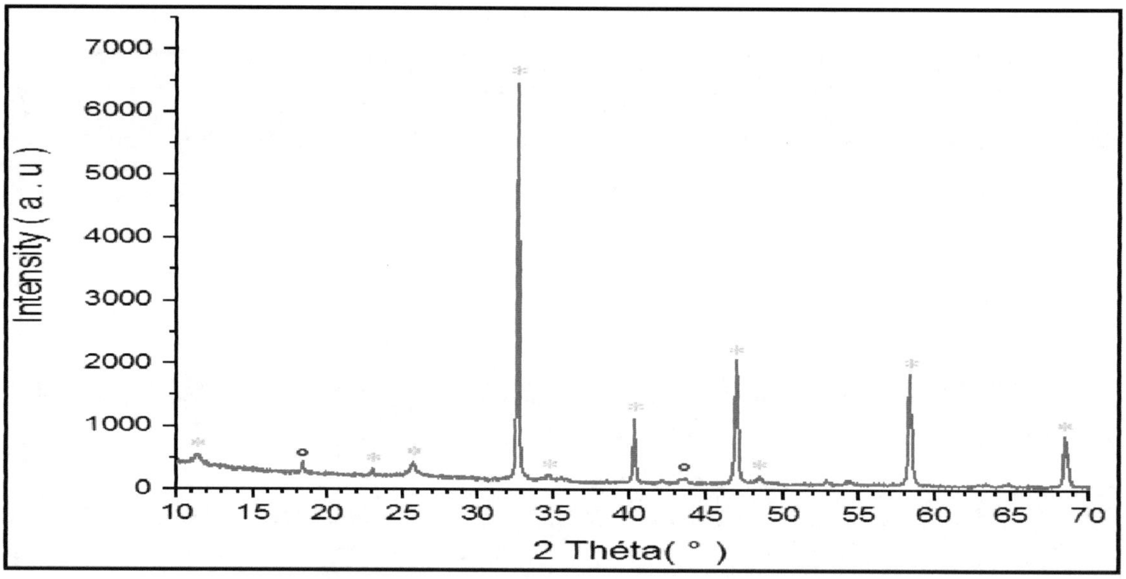

Figure 4. X-ray diffraction of $Li_{0.5}La_{0.5}TiO_3$ obtained by sol–gel method; (*) peaks of LLTO(3*x*=0.5) ; (o) peaks of Li_2TiO_3

Refinement on the compound synthesized at 1000°C (Figure 5) permitted to obtain the cell parameters a=b=3.8703Å and c=7.7529Å using the P4/mmm space group and α=β=γ=90°. Reliability factors are: R_p=39.00%, R_{wp}=26.90%, R_{exp}=17.15% and χ^2=2.46. In table 3, the crystallographic characteristics of $Li_{0.5}La_{0.5}TiO_3$ are presented, and in table 4, the interatomic distances for refined LLTO (3*x*=0.5) are reported, these results are confirmed, for example the distances found in reference [15] for Ti-O are in the range of those of this study.

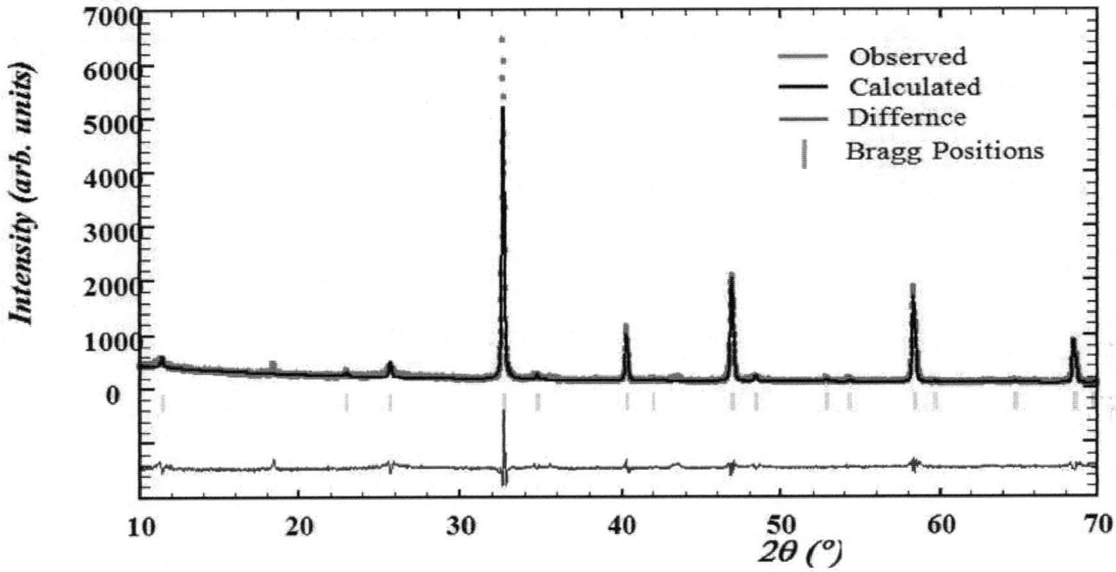

Figure 5. Result of the Rietveld refinement performed on the $Li_{0.5}La_{0.5}TiO_3$ phase heat-treated at 1000°C with calculated and observed diffractograms

Table 3. Wyckoff Sites, Atomic coordinates, for the *P4/mmm* Structure of $Li_{0.5}La_{0.5}TiO_3$

Atom	Site	Atomic coordinates			Occ
		x	y	z	
La1	1a	0.00000	0.00000	0.00000	0.04368
La2	1b	0.00000	0.00000	0.50000	0.02515
Ti	2h	0.50000	0.50000	0.25542	0.13915
O1	1c	0.50000	0.50000	0.00000	0.12169
O2	1d	0.50000	0.50000	0.50000	0.16456
O3	4i	0.00000	0.50000	0.24422	0.34141

Table 4. Interatomic Distances in Å of $Li_{0.5}La_{0.5}TiO_3$

<dTi-O1>	<dTi-O2>	4× <dTi-O3>	4× <dLa1-O1>
1.9803	1.8962	1.9371	2.7367

8× <dLa1-O3>	4× <dLa2-O2>	8× <dLa2-O3>
2.7074	2.7367	2.7708

3.3. Lithium lanthanum titanate for (3x=0.1) $Li_{0.1}La_{0.63}TiO_3$

For samples (3*x*=0.1), the formation of the LLTO phase is produced at 900°C, but with impurities. We have therefore tried to increase the heat treatment temperature to 1000°C in order to obtain pures phases, but unfortunately, we stiff also observe the presence of lanthanum and titanium oxide $La_2Ti_2O_7$ in greater or lesser quantities. The addition of 10% excess of lithium acetate allows us to reduce the quantity of these impurities without eliminating them entirely, an excess of 10% of lithium was added to the stoichiometric mixtures to compensate for lithium loss during processing [16]. (Figure 6).

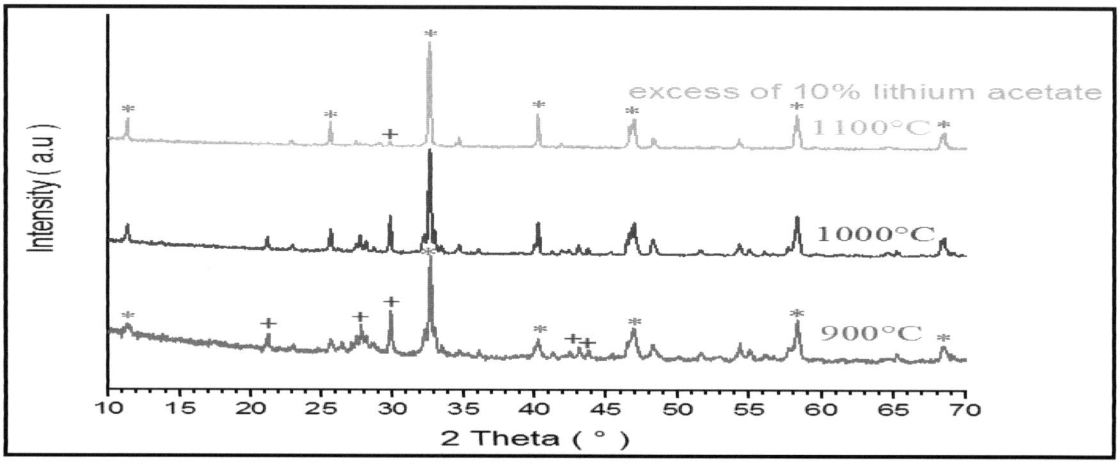

Figure 6. X-ray diffraction of $Li_{0.1}La_{0.63}TiO_3$ obtained by sol–gel method; (*) peaks of LLTO(3*x*=0.1) ; (+) peaks of $La_2Ti_2O_7$

The $Li_{0.1}La_{0.63}TiO_3$ compound heat treated at 1100°C with the 10% excess of lithium acetate was used to perform a Rietveld refinement using the FullProf program (Figure 7). Using the structural model proposed by Ibarra et al [15], we have verified that the LLTO compound (3*x*=0.1), crystallizes well in an Orthorhombic cell (Pmmm; a=3.8743Å; b=3.8632Å and c=7.7816Å , and α=β=γ=90°), and the factors of reliability are: R_p=39.40%, R_{wp}=30.40%, R_{exp}=23.93% and χ^2=1.61. In table 5, the crystallographic characteristics of $Li_{0.1}La_{0.63}TiO_3$ are presented, and in table 6, the interatomic distances for refined LLTO (3*x*=0.1) are reported, these results are in very good agreement with the literature [15].

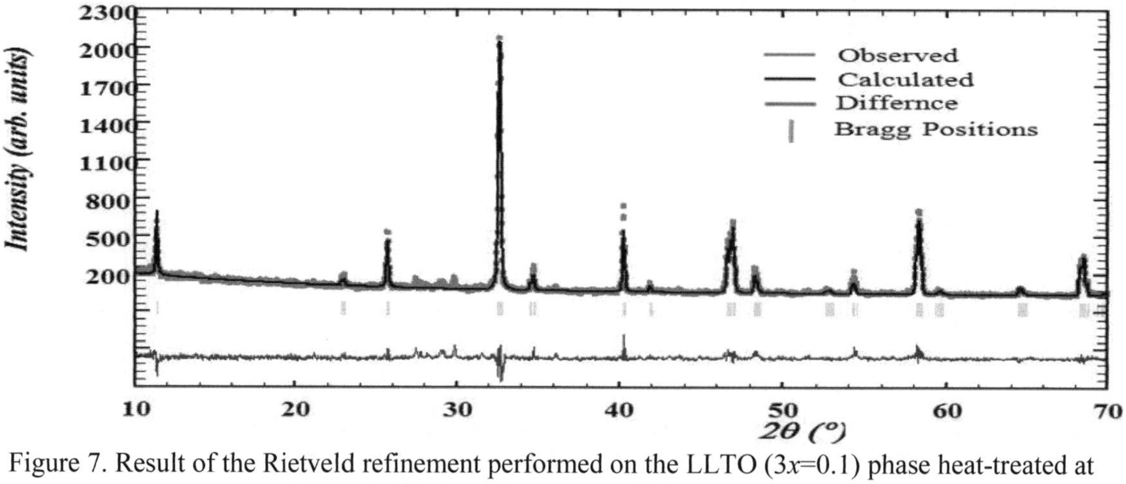

Figure 7. Result of the Rietveld refinement performed on the LLTO (3*x*=0.1) phase heat-treated at 1100°C with the 10% excess of lithium acetate

Table 5. Wyckoff Sites, Atomic coordinates, for the *Pmmm* Structure of $Li_{0.1}La_{0.63}TiO_3$

Atom	Site	Atomic coordinates			Occ
		x	y	z	
La1	1a	0.00000	0.00000	0.00000	0.11467
La2	1c	0.00000	0.00000	0.50000	0.04715
Ti	2t	0.50000	0.50000	0.26272	0.25000
O1	1f	0.50000	0.50000	0.00000	0.12500
O2	1h	0.50000	0.50000	0.50000	0.12500
O3	2r	0.00000	0.50000	0.23426	0.25000
O4	2s	0.50000	0.00000	0.25075	0.25000

Table 6. Interatomic Distances in Å of $Li_{0.1}La_{0.63}TiO_3$

<dTi-O1>	<dTi-O2>	2× <dTi-O3>	2× <dTi-O4>	4× <dLa1-O1>
2.0444	1.8464	1.9498	1.9338	2.7356

4× <dLa1-O3>	4× <dLa1-O4>	4× <dLa2-O2>	4× <dLa2-O3>	4× <dLa2-O4>
2.6559	2.7495	2.7356	2.8297	2.7413

4. Conclusion

The samples of lithium titanate and lanthanum titanate LLTO with ($3x$=0.1), ($3x$=0.3) and ($3x$=0.5); were synthesized by the sol-gel method.
The $Li_{0.3}La_{0.57}TiO_3$ phase was shown to crystallize in the tetragonal structure (space group P4/mmm).
The $Li_{0.5}La_{0.5}TiO_3$ phase has been indexed in a tetragonal cell (P4/mmm space group).
While the $Li_{0.1}La_{0.63}TiO_3$ phase was crystallized in a Orthorhombic cell (Pmmm space group).

References

[1] Trong L D, Thao T T, and Dinh N N. 2015. Characterization of the Li-ionic conductivity of $La_{(2/3-x)}Li_{3x}TiO_3$ ceramics used for all-solid-state batteries. Solid State Ionics. 278: 228–232.
[2] Liu K, Liu Y, Lin D, Pei A, and Cui Y. 2018. Materials for lithium-ion battery safety. Science Advances, 4: eaas9820.
[3] Hu L, and Xu K. 2014. Nonflammable electrolyte enhances battery safety. PNAS. 111(9): 3205–3206.
[4] Abhilash KP, Christopher Selvin P, Nalini B, Somasundaram K , Sivaraj P, Chandra Bose A. 2016. Study of the temperature dependent transport properties in nanocrystalline lithium lanthanum titanate for lithium ion batteries. Journal of Physics and Chemistry of Solids. 91:114–121.
[5] Inaguma Y, Nakashima M. 2013. A rechargeable lithium air battery using a lithium ion conducting lanthanum lithium titanate ceramics as an electrolyte separator. Journal of Power Sources. 228: 250-255.
[6] Stramare S, Thangadurai V, Weppner W. 2003. Lithium Lanthanum Titanates: A Review. Chem. Mater. 15: 3974-3990.

[7] Ban CW, Choi GM. 2001. The effect of sintering on the grain boundary conductivity of lithium lanthanum titanates. Solid State Ionics. 140: 285-292.
[8] Alexander K, Ganesh P, Chi M, Kent P, Sumpter B. 2016. Grain boundary stability and influence on ionic conductivity in a disordered perovskite -a first-principles investigation of lithium lanthanum titanate. MRS Communications. 6:445-463.
[9] Hu X, Quiang W, Huang B. 2017. Preparation and properties of $Li_xLa_{0.5}TiO_3$ perovskite oxide electrolytes. Journal of the American Ceramic Society 100(9): 4153-4158.
[10] Wolfenstinea J, Allena J L, Reada J, Sakamoto J, Gonalez-Doncel G. 2010. Hot-pressed $Li_{0.33}La_{0.57}TiO_3$. Journal of Power Sources. 195:4124-4128.
[11] Bohnke Cl, Regrag B, Berre F L, Fourquet J L, and Randrianantoandro N. 2005. Comparison of pH sensitivity of lithium lanthanum titanate obtained by sol–gel synthesis and solid state chemistry. Solid State Ionics 176: 73-80.
[12] Romero M, Faccio R, Vásquez S, Davyt S, Mombrú A W. 2016. Experimental and theoretical Raman study on the structure and microstructure of $Li_{0.30}La_{0.57}TiO_3$ electrolyte prepared by the sol-gel method in acetic medium. Ceramics International. 0272-8842.
[13] Kitaoka K, Kozuka H, Hashimoto T, Yoko T. 1997. Preparation of $La_{0.5}Li_{0.5}TiO_3$ perovskite thin films by the sol—gel method. Journal of materials. 32-2063-2070.
[14] Fourquet J L, Duroy H, Crosnier-Lopez,. 1996. Structural and Microstructural Studies of the Series $La_{2/3-x}Li_{3x}\square_{1/3-2x}TiO_3$. Solid State Chemistry. 127, 283–294.
[15] Ibarra J, Varez A, Sanz J, Santamaria J, Leon C, Torres-Martinez L M. 2000. Influence of composition on the structure and conductivity of the fast ionic conductors $La_{2/3-x}Li_{3x}TiO_3$ ($0.03 \leq x \leq 0.167$). Solid State Ionics. 134, 219-228.
[16] Ming'en L, Xiaohong Z, Yue J, Jiliang Z. 2016. Comparative study of solid-state reaction and sol-gel process for synthesis of Zr-doped $Li_{0.5}La_{0.5}TiO_3$ solid electrolytes. Ionics. 22, 2151-2156.

Diffuse and relaxor phase transitions of $Ba_{0.95}Bi_{0.05}Ti_{1-x}Fe_xO_3$ ceramics at x=0.00 to 1.00 of Fe content, prepared by solid state method.

GOUITAA Najwa*, LAMCHARFI Taj-dine, ABDI Farid, ECHATOUI Nor-Said and AMARASS Mustapha.

Signals, Systems and Components Laboratory (LSSC), Electrical Engineering Department, University Sidi Mohamed Ben Abdellah USMBA, FST. Fez, Imouzzer Road B.P. 2202, Morocco.

author's e-mail address: najwa.gouitaa@gmail.com

Abstract:

The dielectric properties of iron and bismuth co-substituted $BaTiO_3$ ceramic with the formula: $Ba_{0.95}Bi_{0.05}Ti_{1-x}Fe_xO_3$ ($BBTF_x$) for x=0.00 to 1.00, synthesis with solid state route, were characterized. The results of dielectric permittivity dependence of temperature show three dielectric phase transitions from a rhombohedral ferroelectric to orthorhombic ferroelectric (T_{R-O}) then to a tetragonal ferroelectric phase (at T_{O-T}), and finally to cubic paraelectric at the Curie temperature (T_C). Calculating the relaxation parameters of ΔT_{relax} and the dispersion factor $\Delta\varepsilon'_{r,max} / \varepsilon'_{r,max}$, we have found that all the ceramics present a relaxation and a dispersion phenomenon. And the phase transition T_{O-T} for x = 0.00 has the highest value of ΔT_{relax} while the dispersion factor has the highest value for the phase transition T_m at x = 0.60. The diffuseness parameter γ refined from Uchino law showed a diffuse phase transition for all the ceramics. While γ has a value close to 1 for T_{R-O} at x = 0.00, which indicate a standard Curie-Weiss law. The dielectric evolution as function of frequency shows a resonance behavior for x=0.00 of Fe content while for the other ceramics we obtained a classical ferroelectric behavior.

Keywords:

Dielectric, $BaTiO_3$, phase transition, relaxation, dispersion, diffuse, resonance.

I. Introduction

For electronic industry, the $BaTiO_3$ based material has been widely used and has an important role in the applications of actuators, sensors, resonators, ultrasonic transducers, bolometers, solid state refrigeration and other capacitive applications [1-3]. The basic properties of ferroelectric ceramics based capacitors are the high dielectric constant, low dielectric loss, and high hysteresis free strain near room temperature over the wide frequencies. The dielectric and ferroelectric properties of BT-based can be modified by a large variety of substitutions possible at Ba-site or Ti-site independently or simultaneously. In order to improve their dielectric and ferroelectric properties these substituents can be isovalent or heterovalent ions. Effects of isovalent substitutions such as Pb^{2+}, Ca^{2+}or $Sr^{2+}$for Ba-site and Zr^{4+}, Hf^{4+} or Mn^{4+} for Ti^{4+}-site on microstructure, dielectric properties of BT-based ceramics have already been investigated [4-6]. Whereas the rare-earth substitution of heterovalent ions such as Dy^{3+}, Tb^{3+}, Eu^{3+}, Nd^{3+}, Gd^{3+}, Yb^{3+}, Sm^{3+}, Er^{3+}, Ho^{3+}, Sb^{3+}, Sc^{3+}, La^{3+}, Nb^{5+}, Bi^{3+} for Band Ti-site cause charge imbalance and creation of vacancies on A sites or B sites [7-12].

$BiFeO_3$-ABO_3 systems have attracted great attention as a means to rise structural stability and sinterability [13,14]. Processing of $BiFeO_3$ with other perovskite structured materials, such as $BaTiO_3$, would prevent the formation of secondary phases [15,16] and have excellent ferroelectric properties (T_C= 120 °C).

Recently, several researchers reported multiferroic perovskite ceramics which have antiferromagnetic – ferroelectric and antiferromagnetic /weak ferromagnetic ferroelectric systems [17,18]. As a typical ferroelectro-magnetic material, the $BiFeO_3$ has been a member of rare earth pervoskite family having semi-conducting characteristics at room temperature with abnormal dielectric constant and loss which can be a useful substitution for BT. As there are few extensive reports on the simultaneous substitution of $Bi^{3+}$and Fe^{3+} in $BaTiO_3$. Therefore, we have studied the structural and multiphase phase transitions of $Ba_{0.95}Bi_{0.05}Ti_{1-x}Fe_xO_3$ ceramics for x=0.00 to 1.00, in our previous work [19]. We have found a coexistence of three phases which are tetragonal, hexagonal and pseudo-cubic in a large marge of concentration of Fe (from x=0.40 to 0.80). And The results of dielectric properties show three dielectric relaxation phase transitions from a rhombohedral ferroelectric to orthorhombic ferroelectric (T_{R-O}) then to a tetragonal ferroelectric phase (at T_{O-T}), and finally to cubic paraelectric at the Curie temperature (T_C). The present paper, is focusing on the detailed study of these phase transitions by identifying a relaxation and diffuseness parameters to describe these important phenomena. After that, we have studied the dielectric permittivity and the dielectric loss of these ceramics as function of frequency at different temperatures.

II. Experimental

The $Ba_{0.95}Bi_{0.05}Ti_{1-x}Fe_xO_3$ ($BBTF_x$) ceramics for x=0.00 to 1.00 were prepared by the conventional solid-state method using oxides of $BaCO_3$, Bi_2O_3, TiO_2 and $Fe_2O_3$with high-purity of 99%, 99%, 99% and 97% respectively and from Sigma-Aldrich provider. The doping site of Fe and Bi ions in $BaTiO_3$ was weighted by stoichiometric proportions of raw materials according to the reaction:

$$0.95\ BaCO_3 + \tfrac{1}{2}\ 0.05\ Bi_2O_3 + \tfrac{1}{2}\ x\ Fe_2O_3 + (1 - \frac{0.05}{4} - x)\ TiO_2 \rightarrow Ba_{0.95}Bi_{0.05}Ti_{1-0.05/4-x}Fe_xO_{3-x/2} + CO_2$$

The starting powders were weighted in stoichiometric proportion and grinded for 30 min after that milled under acetone for 4h. The obtained powders were dried at 80°C for 24h. The dried powders were mixed using agate mortar for 30 min and then de-carbonated and pre-reacted by calcining in alumina at 1100 °C for 4h. After calcination, the powders were mixed for 30 min.

The calcined powders were pressed into pellets, after adding few drops of 1 wt% Polyvinyl Alcohol (PVA) as a binder, and sintered in air at 1200 °C for 6h. The dielectric properties as function of frequency and temperature were studied with Agilent E4980A (20Hz-2MHz).

III. Results and discussion

1. Dielectric properties as function of temperature

The figure.1 shows the evolution of the dielectric permittivity, as a function of temperature from the room temperature to 600 °C and at different frequencies (from 5 KHz to 2 MHz), for $BBTF_x$ samples (x = 0.00, 0.20, 0.40, 0.60, 0.80 and 1.00). The spectra show three dielectric anomalies at different temperatures. Based on our previous work. The first maximum at low temperature~ 150 °C, corresponds to a phase transition from the ferroelectric rhombohedral phase to the ferroelectric orthorhombic phase T_{R-O}. The second maximum at intermediate temperatures (~500 °C) corresponds to a phase transition from the ferroelectric orthorhombic phase to the ferroelectric tetragonal phase T_{O-T}. While the third maximum at high temperatures (~ 600 °C) corresponds to a phase transition from the tetragonal ferroelectric phase to the cubic paraelectric phase T_m.

These three phase transitions show a dielectric relaxation phenomenon in all these ceramics and a diffuse behavior.

1.1. Relaxation phenomenon

In order to describe the relaxation behavior of the phase transitions, three regions are distinguished: region I corresponding to the first phase transition T_{R-O}, region II in which the second peak persists T_{O-T} and region III for high temperatures where the paraelectric-ferroelectric phase transition T_m takes place.

The adjustment parameters and the characteristics of the relaxer phenomenon for the three phase transitions are summarized in the table. 1, with:

ΔT_{relax} is the relaxation parameter at T_m and it is defined by:

$$\Delta T_{relax}\ (°C) = T_m\ (100\ KHz) - T_m\ (5\ KHz)$$

And $\Delta\varepsilon'_{r,max}$ / $\varepsilon'_{r,max}$ dispersion factor defined by:

$$\Delta\varepsilon'_{r,max}/\varepsilon'_{r,max} = \frac{\varepsilon'_r\ (5\ KHz) - \varepsilon'_r\ (100\ KHz)}{\varepsilon'_r\ (5\ KHz)}$$

The values of these parameters are grouped in table.1. We note that the values of ΔT_{relax} increase with the increase in the Fe level for the first phase transition T_{R-O} and decrease with the increase in Fe for the other two phase transitions T_{O-T} and T_m. The phase transition T_{O-T} for x = 0.00 has the highest value of ΔT_{relax}. While the dispersion factor has the highest value for the phase transition T_m at x = 0.60.

Table.1. Parameters value of the relaxer phenomenon ΔT_{relax} and $\Delta\varepsilon'_{r,max}/\varepsilon'_{r,max}$ for the three phase transitions.

Phase transition		x	ΔT_{relax} (°C)	$\Delta\varepsilon'_{r,max}/\varepsilon'_{r,max}$
T_{R-O}		**0.00**	3	0.16
		0.20	7	0.18
		0.40	5	0.73
		0.60	9	0.5
		0.80	17	0.37
		1.00	26	0.7
T_{O-T}		**0.00**	56	0.67
		0.20	30	0.056
		0.40	17	0.51
		0.60	21	0.6
		0.80	3	0.64
		1.00	8	0.55
T_m		**0.00**	----	----
		0.20	10	0.52
		0.40	9	0.63
		0.60	4	0.86
		0.80	7	0.60
		1.00	7	0.72

Joby E. Joy et al. [20] reported that the dielectric properties of the perovskite oxides of formula $Ba_3M^{III}TiM^{V}O_9$ for M^{III} = Fe, Ga, Y and Lu, M^{V} = Nb, Ta and Sb showed a relaxer-like behavior only for the M^{III} = Fe oxides, whereas all the other oxides showed a classical dielectric behavior. In our case, the relaxation is also present for BBTO not substituted by Fe ion, which suggests that this phenomenon is due to the presence of Bi^{3+} in the Ba^{2+} site corresponding to M^{III}. However, the BT material doped with Fe at 0.20, although it does not correspond to an M^{V} ion, the material is a relaxer which indicates that the combination of the two substitutions (Bi and Fe) favors the relaxer phenomenon. The co-substitution of these two ions (Bi^{3+} and Fe^{3+}) promotes the relaxer character.

On the other hand, it has been suggested that the defects such as oxygen vacancies are responsible for dielectric relaxation in the high temperature region. The existence of oxygen vacancies is characteristic of $BaTiO_3$ ferroelectric ceramics prepared by high temperature sintering. In $BaTiO_3$, oxygen vacancies with n-type carriers are generated by the following reaction:

$$O_O \leftrightarrow V\ddot{O} + 2\,e' + \tfrac{1}{2}\,O_2(g)$$

The electrons released in this reaction can be sensed by the Fe^{3+} ions to form Fe^{2+} ions:

$$Fe^{3+} + e' \leftrightarrow Fe^{2+}$$

Where O_o is the oxygen ion in the oxygen lattice site, with a neutral charge, $V\ddot{O}$ denotes the oxygen gap in the $BaTiO_3$ lattice with an effective charge of +2 according to the Kräger-Vink notation. The change in ionization can be written by the formula:

$$V\ddot{O} + e' \leftrightarrow V\dot{O}$$

With: $V\dot{O}$ denotes the oxygen gap in the $BaTiO_3$ with an effective charge of +1.

The oxygen vacancies $V\ddot{O}$ and $V\dot{O}$ would be attracted by the negatively charged impurity center Fe'_{Ti} (Fe ion in a Ti lattice site, with -1 charge -Fe^{3+}) Fe''_{Ti} (Fe ion in a Ti lattice site, with -2 charge-Fe^{2+}) creating the defective complexes [21] according to the formula:

$$Fe'_{Ti} -- V\dot{O} \quad \text{et} \quad Fe''_{Ti} -- V\ddot{O}$$

The coexistence of Fe^{3+} and Fe^{2+} causes an electron jump between the Fe^{3+} and Fe^{2+} ions, with the oxygen vacancies acting as a bridge between them. In addition, it can lead to increased values of electrical conductivity, dielectric losses, and space charge build-up at grain boundaries [22].

For all these compounds, a relaxer behavior was detected. Thus, the replacement of Ti^{4+} by Fe^{3+} and of Ba^{2+} by Bi^{3+} have the effect of promoting the relaxation behavior for the three transition temperatures with values which vary according to the rate of Fe^{3+} substitution.

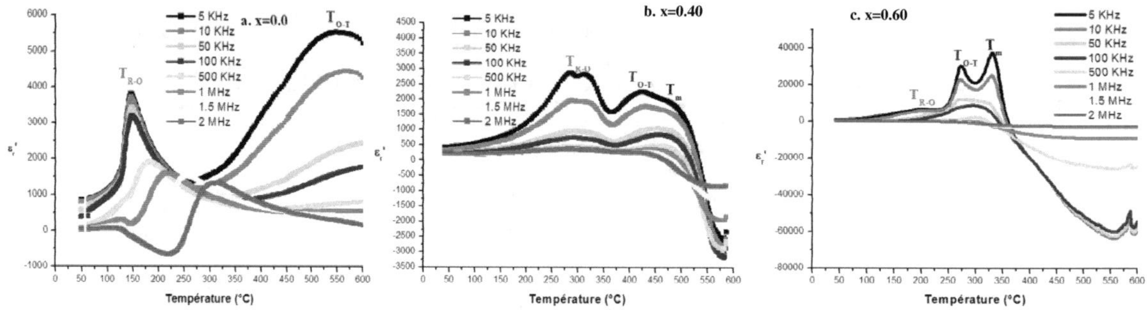

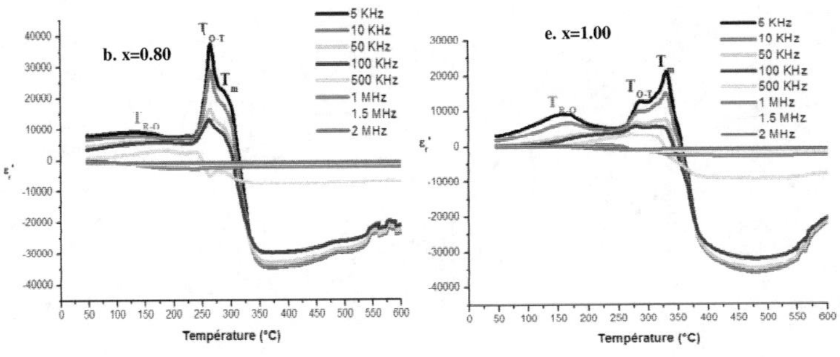

Figure.1. Temperature dependence of dielectric permittivity of $Ba_{0.95}Bi_{0.05}Ti_{1-x}Fe_xO_3$ ceramics for x=0.00 to 1.00.

1.2. Diffuse character

1.2.1. Curie - Weiss law

The normal ferroelectrics are characterized by a steep phase transition, at T_C, from the polar to the non-polar phase, reflected by the discontinuity of ε_r above T_C. The transition temperature T_C is independent of the frequency and the paraelectric behavior of a normal ferroelectric is described by the Curie - Weiss law:

$$\varepsilon'_r = \frac{C}{T-T0}$$

Where C denotes the Curie Weiss constant and T_0 the Curie Weiss temperature.

To study this type of behavior (ferroelectric with diffuse phase transition), we have plotted the curves of $1/\varepsilon_r$ as a function of temperature for the three phase transitions as it is shown in the figures. 2, 3and 4. These curves can be divided into three regions: the first region represents the ferroelectric behavior up to the transition of T_m, the second region close to the transition indicates a diffuse transition from T_m to the temperature T_{cw} and the third region, above T_{cw}, represents the linear behavior of $1 / \varepsilon'_r$ as a function of temperature according to the Curie Weiss diagram. The Curie Weiss temperature T_o is obtained by extrapolating the third region to the temperature axis. And the factors Ω and C represent respectively the inverse of the slopes of the line $1 / \varepsilon'_r(T)$. These different constants are listed in the table. 2. The values of the ratio Ω / C obtained are close to the -2 value for the $T_{R\text{-}O}$, $T_{O\text{-}T}$ transitions at x = 0.20, x = 0.00 and for T_m at x = 0.20 and 0.80, which corresponds to an ideal transition second order at these doping rates and close to -8 for the transition T_m at x = 0.40, 0.60 and 1.00 which corresponds to a first order transition [23,24]. While for the other substitution rates the ratio Ω / C is very far from -2 and -8 because the phase transitions at these substitution rates are very diffuse, which limits the application of the Curie-Weiss law.

The parameter ΔT_m, is used to indicate the degree of deviation from the Curie Weiss law and it is defined as follows:

$$\Delta T_m = T_{CW} - T_m$$

Where T_{CW} denotes the temperature at which the permittivity begins to deviate from Curie Weiss law and T_m represents the temperature of the minimum of $1 / \varepsilon'_r$.

The values of this parameter (Table. 2) are all different from zero. So, we have a deviation from the Curie Weiss behavior for all the ceramics, which has been qualified for a short-range association between the nanopolar domains [25]. In addition, the deviation from the Curie Weiss law at some temperatures range above T_m is one of the most features of ferroelectric relaxers [26].

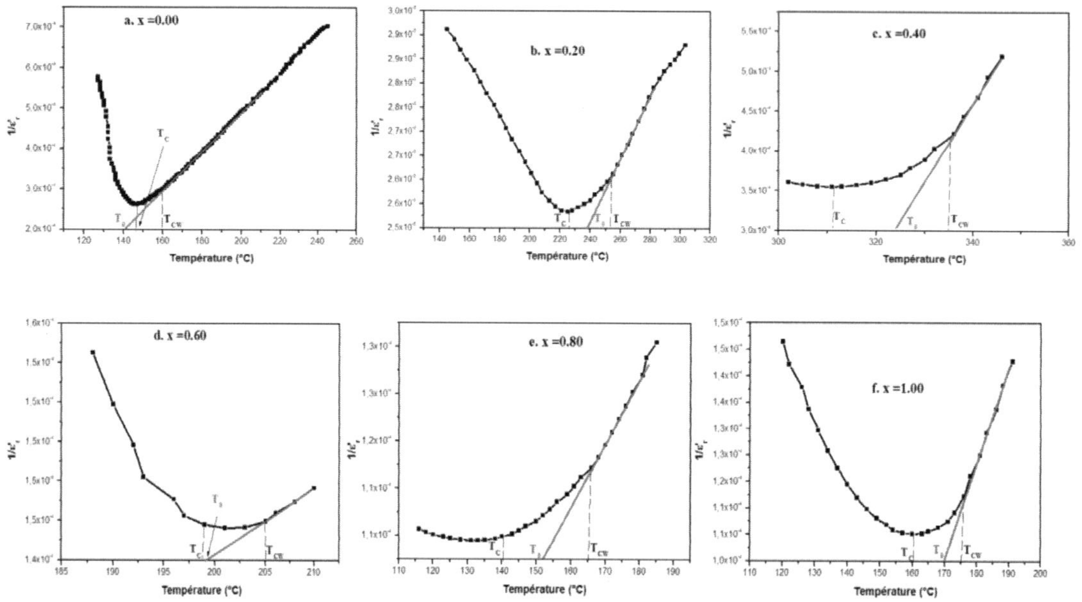

Figure. 2 Evolution of 1/ε'$_r$ as function temperature at 5 kHz for $T_{R\text{-}O}$ phase transition for $BBTF_x$ at x= a. 0.00, b.0.20, c.0.40, d.0.60, e.0.80 and f.1.00.

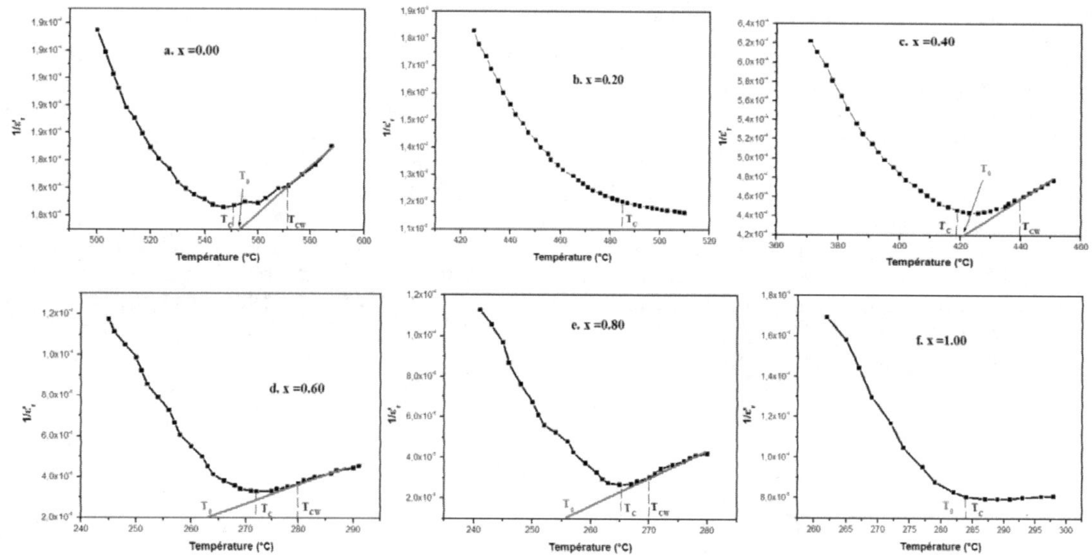

Figure. 3 Evolution of 1/ε'$_r$ as function temperature at 5 kHz for $T_{O\text{-}T}$ phase transition for $BBTF_x$ at x= a. 0.00, b.0.20, c.0.40, d.0.60, e.0.80 and f.1.00.

IOP Conf. Series: Materials Science and Engineering 1160 (2021) 012006 doi:10.1088/1757-899X/1160/1/012006

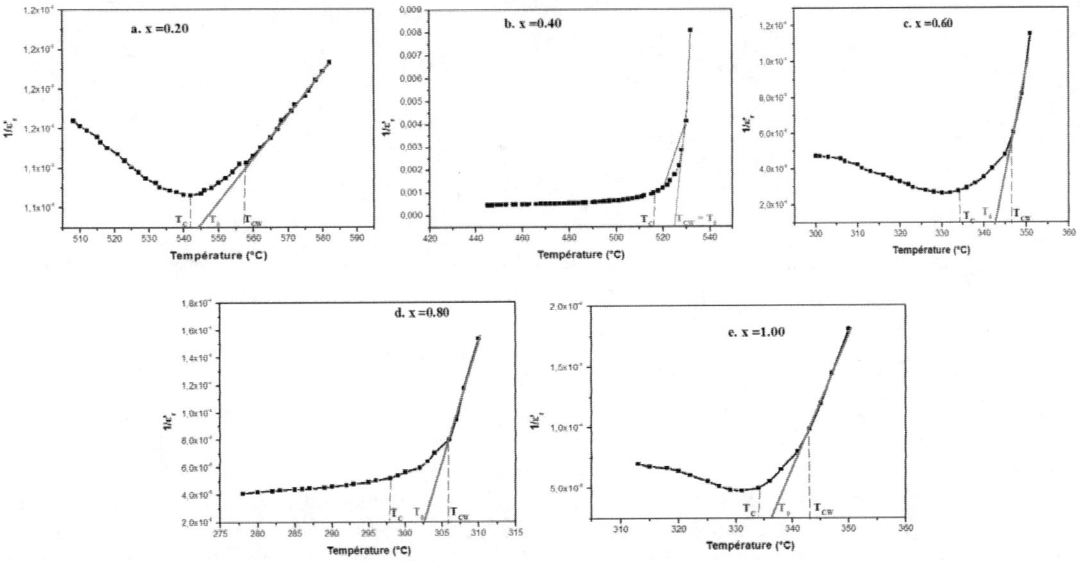

Figure. 4 Evolution of $1/\varepsilon'_r$ as function temperature at 5 kHz for T_m phase transition for $BBTF_x$ at x= a. 0.20, b.0.40, c.0.60, d.0.80 and e.1.00.

Table.2. Curie Weiss parameters for the three phase transitions.

Phase transition	x	ΔT_m (°C)	C (10^5 °C)	Ω (10^5 °C)	Ω/ C
T_{R-O}	**0.00**	13	4.3	-0.45	-0.10
	0.20	27	1	-1.7	-1.7
	0.40	24	7.2	-1.42	-0.2
	0.60	6	25	-10	-0.4
	0.80	13	13.3	-1.5	-0.11
	1.00	5	5	-0.74	-0.15
T_{O-T}	**0.00**	19	9.5	-23	-2.42
	0.20	18	1.01	-0.13	-0.13
	0.40	19	5	-4.27	-0.85
	0.60	17	10.3	-2.30	-0.22
	0.80	15	7.35	-2.32	-0.32
	1.00	6	1.1	-0.74	-0.67
	0.00	----	----	---	--
	0.20	16	4.5	-5.5	-1.22
	0.40	6	0.1	-1	-10

T_m	**0.60**	5	1.04	-11.76	-11.31
	0.80	3	0.5	-1.02	-2.04
	1.00	7	0.95	-8	-842

1.2.2. Uchino law

Figures 5, 6 and 7 illustrate the ferroelectric behavior of $BBTF_x$ compounds (for x = 0.00 to 1.00) according to the modified Uchino law [27] , in a straight line form the equation:

$$\mathrm{Ln}\,[((\varepsilon'_{r,max} / \varepsilon'_{r})-1)*2] = \gamma \ln (T-T_m) - \gamma \ln\delta$$

With: δ Gaussian diffusion factor (K) and γ indicate the character of the transition: γ = 1 for an ideal Curie transition, γ = 2 for the relaxer behavior.

Linear relationships are observed in the plot of Ln $[((\varepsilon'_{r,max} / \varepsilon'_{r}) -1)]$ versus Ln $(T-T_m)$ at 5 kHz of frequency for the three regions. By fitting the experimental data to the Uchino equation, the values of γ and δ are deduced and are presented in the table. 3. We note that for some ceramics and phase transition, the values of γ are in the range 1.35 to 1.90, which corresponds to an incomplete diffuse phase transition. The diffusivity γ has a value close to 1 for $T_{R\text{-}O}$ at x = 0.00 (BBTO not co-substituted) which indicate a standard Curie-Weiss law. While for other ceramics at x = 0.20 and 1.00 for $T_{O\text{-}T}$ as well as at x = 0.00 and 0.80 for T_m, Uchino's law is no longer applicable because the Curie temperature distribution at these doping rates does not follow the Gaussian distribution as it was described by Smolenski [28].

The diffuse phase transition behavior of these samples can be explained by the presence of micro-heterogeneities in these materials due to the thermochemical diffusion controlled by solid state reactions. The micro-heterogeneities are due to the random occupation of B and A sites by different ions. Such a heterogeneous distribution of cations leads to different states of polarization and therefore to different relaxation times in different regions [29].

On the other hand, basing on the values of the diffusion factor δ, it can be seen that when the Fe level increases this factor varies irregularly for all the phase transitions.

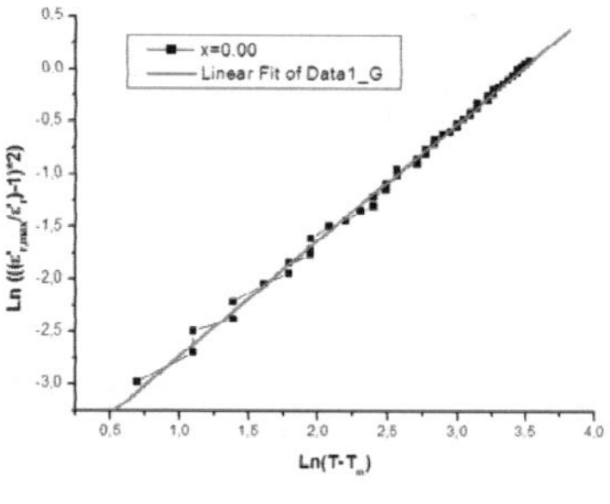

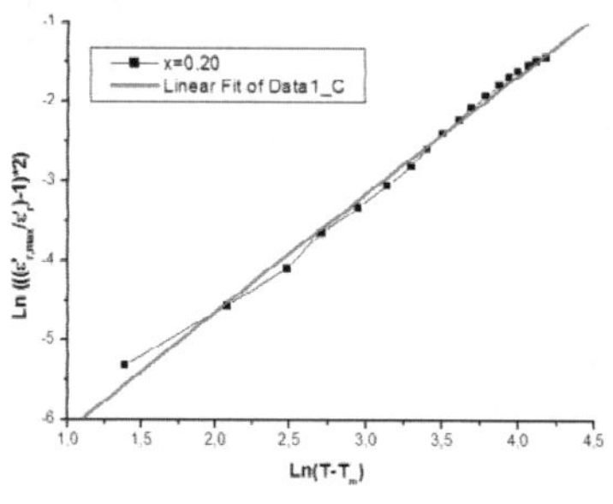

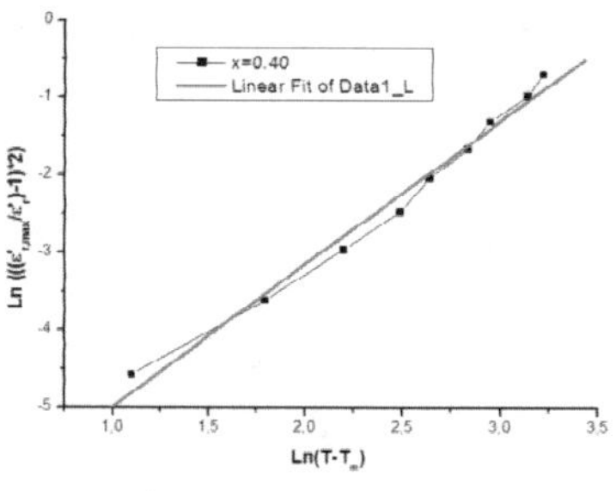

IOP Conf. Series: Materials Science and Engineering 1160 (2021) 012006 doi:10.1088/1757-899X/1160/1/012006

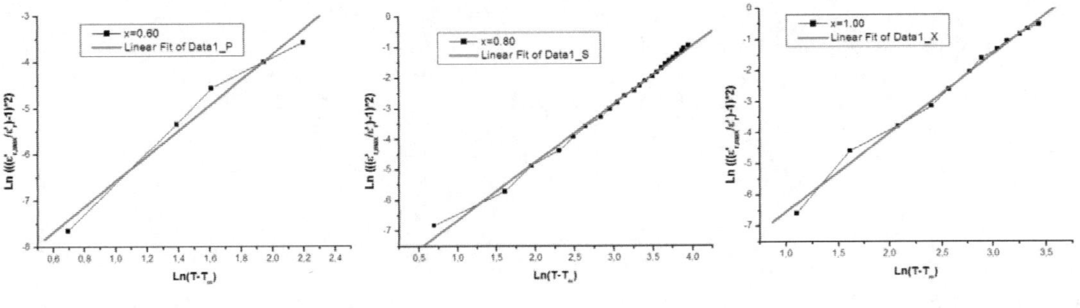

Figure. 5. Ln [((ε'$_{r,max}$ / ε'$_r$) -1)] versus Ln (T-T$_m$) at 5 kHz for T$_{R-O}$ phase transition for BBTF$_x$ ceramics at x= 0.00 to 1.00.

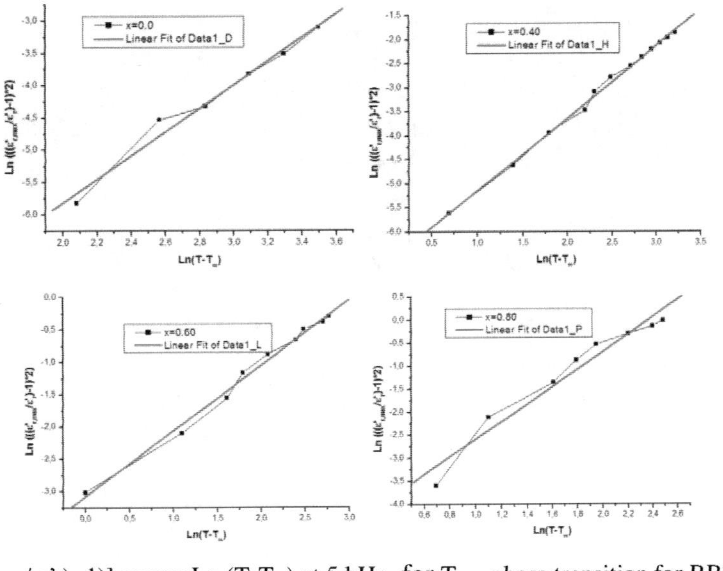

Figure. 6. Ln [((ε'$_{r,max}$ / ε'$_r$) -1)] versus Ln (T-T$_m$) at 5 kHz for T$_{O-T}$ phase transition for BBTF$_x$ ceramics at x= 0.00, 0.40, 0.60 and 0.80.

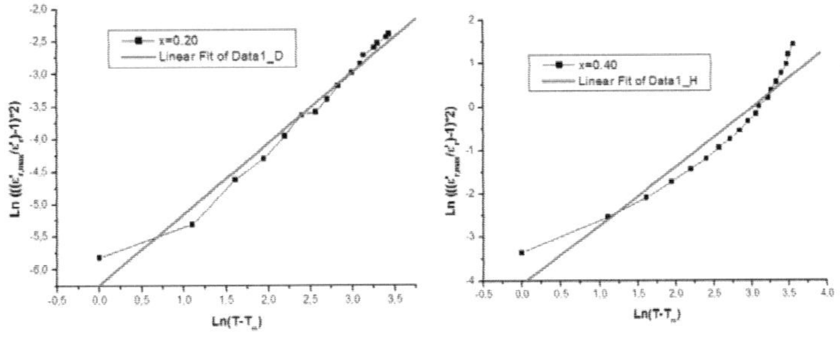

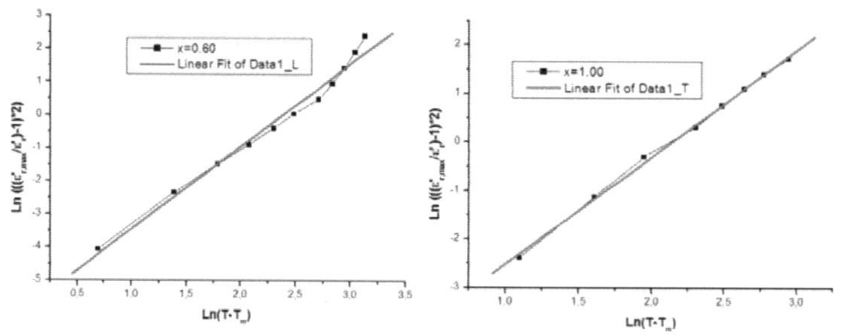

Figure. 7. Ln [((ε'$_{r,max}$ / ε'$_r$) -1)] versus Ln (T-T_m) at 5 kHz for T_m phase transition for $BBTF_x$ ceramics at x= 0.20, 0.40, 0.60 and 1.00.

Tableau. 3. γ and δ values for the tree phase transitions at different Fe contents.

Phase transition	T_{R-O}						T_{O-T}						T_m					
Fe content	0.0	0.2	0.4	0.6	0.8	1.0	0.0	0.2	0.4	0.6	0.8	1.0	0.0	0.2	0.4	0.6	0.8	1.0
γ	1.1	1.49	1.84	>2	1.89	>2	1.82	----	1.51	1.01	1.90	----	----	1.08	1.35	>2	----	>2
δ	33.1	164	40.44	30	90	36.6	181	----	84.8	20.9	10.7	----	----	324	20.3	10.7	----	8.5

2. Dielectric properties as function of frequency

The figure. 8 shows the dielectric constant at different temperatures as a function of frequency of all ceramics in the frequency range from 20 Hz to 2 MHz. For the BT substituted at 0.05 of Bi (x = 0.00), the permittivity curves show a resonance phenomenon and the peak shifts towards low frequencies, for temperatures below 160 ° C, then towards high frequencies beyond this temperature. This resonance phenomenon is no longer visible for temperatures above 220 ° C.

While the $BBTF_x$ ceramics with x = 0.20 to 1.00, show a classical ferroelectric behavior. In fact, the values of the dielectric constant decrease rapidly with increasing frequency and become almost independent of frequency. This decrease in the dielectric constant with the increase in frequency can be explained by the phenomenon of dipole relaxation due to the contribution of interfacial polarization of the Maxwell - Wagner type [30]. The dipoles with large effective masses (eg oxygen vacancies) may follow the applied field at low frequency and does not follow it at higher frequencies.

The figure. 9 shows the variation of dielectric loss (tan δ) as a function of frequency at different temperatures. The curves of x = 0.00 (figure. 9.a) show that the dielectric loss decreases with the frequency for the low frequencies then becomes almost constant and then reaches a maximum for the temperatures of 160 ° C and 220 ° C. These maxima corresponding to the resonance frequencies of ε'$_r$ which move towards the high frequencies and decrease in intensity with the increase of the temperature then they are no longer observable for the other temperatures and this is in relation with the phenomenon of resonance which is no longer observed because we exceed the frequency range of the device.

For BBTO ceramics substituted with Fe for x = 0.20 to 1.00, the dielectric losses have the same evolution. In fact, the latter decrease rapidly at low frequencies then remain almost constant for the other frequencies, which indicate the normal ferroelectrics behavior. In addition, the decrease in tan δ at low frequencies is attributed to the fact that the jump frequency of the charge carriers cannot follow the variations of the electric field applied beyond a certain frequency limit.

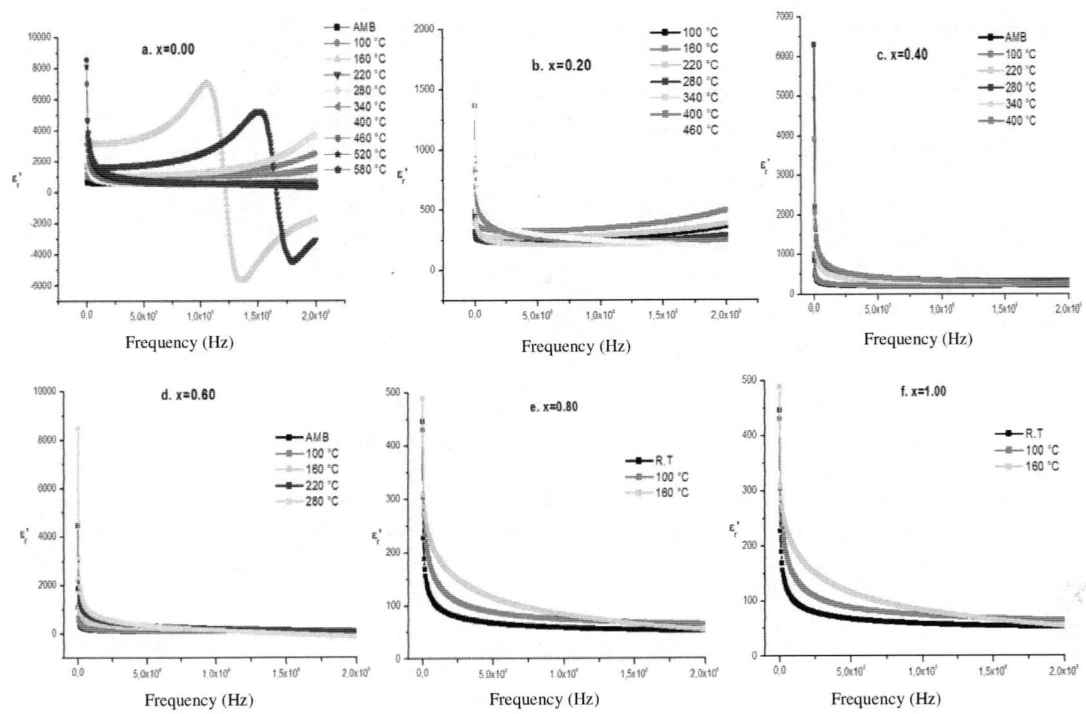

Figure. 8. Evolution of dielectric permittivity as function of frequency for $BBTF_x$ ceramics at x= a.0.00, b. 0.20, c.0.40, d.0.60, e.0.80 and f.1.00.

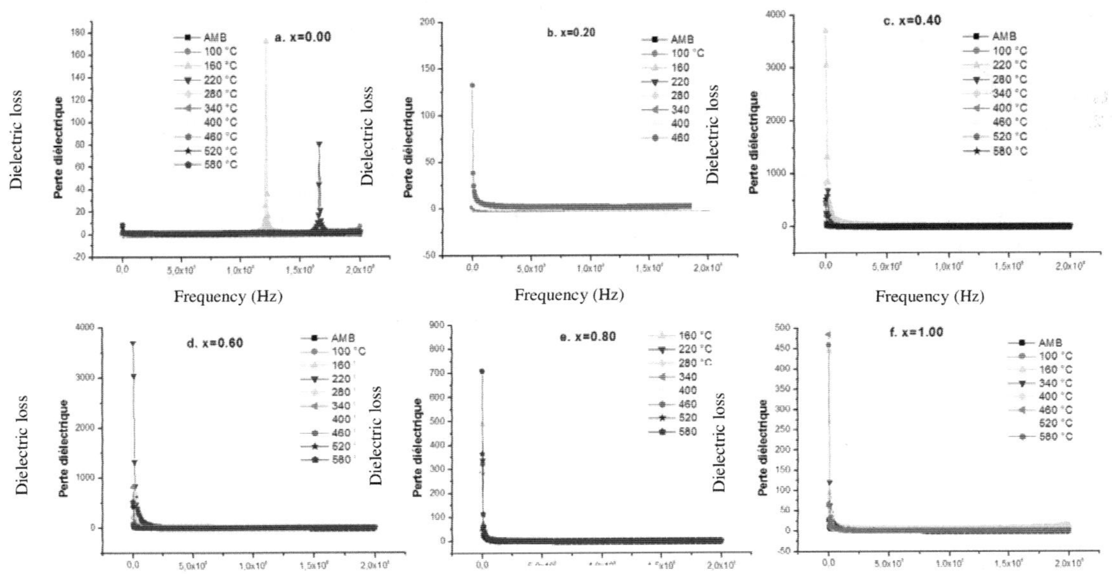

Figure. 9 Evolution of dielectric loss as function of frequency for $BBTF_x$ for x= a.0.00, b. 0.20, c.0.40, d.0.60, e.0.80 and f.1.00.

IV. Conclusion

In this work we have study the dielectric properties of $Ba_{0.95}Bi_{0.05}Ti_{1-x}Fe_xO_3$ ceramics as function of temperature and frequency. The results show the existence of three phase transitions $T_{R\text{-}O}$, $T_{O\text{-}T}$ and T_m which exhibit a relaxation behavior with a diffuse phenomenon. The relaxation parameters ΔT_{relax} and the dispersion factor $\Delta\varepsilon'_{r,max}$ / $\varepsilon'_{r,max}$ calculated show that the phase transition $T_{O\text{-}T}$ for $x = 0.00$ has the highest value of ΔT_{relax} while the dispersion factor has the highest value for the phase transition T_m at $x = 0.60$. The diffuseness parameter γ refined from modified Uchino law showed a diffuse phase transition for all the ceramics. While γ has a value close to 1 for $T_{R\text{-}O}$ at $x = 0.00$, which indicate a standard Curie-Weiss law.

Conflict of Interest
The authors declare that they have no conflict of interest.

References

[1] J.E. Sundeen, R.C. Buchanan **2001**. Thermal sensor properties of cermet resistor films on silicon substrates. Sens. Actuators A, 2875, pp. 1-7.
[2] L. Wang, J. Wang, B. Li, X. Zhong, F. Wag, H. Song, Y. Zeng, D. Huang, Y. Zhou **2014.** Enhanced room temperature Electrocaloric effect in barium titanate thin films with diffuse phase transition. RSC Adv., 4 p. 21826.
[3] Y. Bai, X. Han, X. Cheng Zheng, L. Qiao **2013**. Both high reliability and giant Electrocaloric strength in $BaTiO_3$ ceramics. Sci. Rep., 3, p. 2895.
[4] O.P. Thakur, C. Prakash **2003**. Dielectric properties of samarium substituted barium strontium titanate. Phase Transit., 76 , pp. 567-574.
[5] Z.C. Li, H. Zhang, X.D. Zou **2005**. Synthesis of Sm-doped $BaTiO_3$ ceramics and characterization of a secondary phase. Mat. Sci. Eng. B, 116, pp. 34-39.
[6] Y. Chen, X.L. Dong, R.H. Liang, et al **2005**. Dielectric properties of $Ba_{0.6}Sr_{0.4}TiO_3/Mg_2SiO_4/MgO$ composite ceramics. J. Appl. Phys., 98, p. 064107.
[7] G.P. Du, Z.J. Hu, Q. F. Han, X.M. Qin **2010**. Effects of niobium donor doping on the phase structures and magnetic properties of Fe-doped $BaTiO_3$ ceramics. J. Appl. Phys., 492, pp. 79-82.
[8] X.J. Chou, J.W. Zhai, H.T. Jiang **2007**. Dielectric properties and relax or behavior of rare-earth (La, Sm, Eu, Dy, Y) substituted barium zirconium titanate ceramics. J. Appl. Phys., 102, p. 084106.
[9] S. Devi, A.K. Jha **2009**. Structural, dielectric and ferroelectric Properties of tungsten substituted barium titanate ceramics. Asian J. Chem., 21, pp. S117-S124.
[10] J. Li, D.R. Jin, L.X. Zhou **2012**. Dielectric properties of barium strontium titanate (BST) ceramics synthesized by using mixed-phase powders calcined at varied temperatures. Mater. Lett., 76, pp. 100-102.
[11] W. Cai, C.L. Fu, J.C. Gao **2009**. Effects of grain size on domain structure and ferroelectric properties of barium zirconatetitanate ceramics. J. Alloys Compd., 480, pp. 870-873.
[12] M.R. Panigrahi, S. Panigrahi **2009**. Synthesis and microstructure of Ca-doped $BaTiO_3$ ceramics prepared by high-energy ball-milling. Physica, B404, pp. 4267-4272.
[13] L. Jin, F. Li and S. Zhang **2014**. Decoding the fingerprint of ferroelectric loops: Comprehension of the material properties and structures. J. Am. Ceram. Soc. 97, 27.
[14] L. Jin, R. Huo, R. Guo, F. Li, D. Wang, Y. Tian, Q. Hu, X. Wei, Z. He and Y. Yan **2016.** Diffuse phase transitions and giant electrostrictive coefficients in lead-free Fe^{3+}-doped 0.5 Ba $(Zr_{0.2}Ti_{0.8})$ $O_{3\text{-}0.5}$ $(Ba_{0.7}Ca_{0.3})$ TiO_3 ferroelectric ceramics, ACS Appl. Mater. Interfaces. 8, 31119.
[15] M. Mahesh Kumar, A. Srinivas and S. Suryanarayana **2000**. Structure property relations in $BiFeO_3/BaTiO_3$ solid solutions, J. Appl. Phys. 87, 855.
[16] M. T. Buscaglia, L. Mitoseriu, V. Buscaglia, I. Pallecchi, M. Viviani, P. Nanni and A. S. Siri **2006.** Preparation and characterization of the magneto-electric x $BiFeO_3$ - (1-x) $BaTiO_3$ ceramics, J. Eur. Ceram. Soc. 26, 3027.
[17] S. Kim, C. I. Cheon, P. W. Jang, Y. N. Choi and C. H. Lee **2004.** Ferroelectric and ferromagnetic properties of 0.2 $BiFeO_3$–0.2$RFeO_{3\text{-}0.6}$ $ATiO_3$(R = Pr, Nd and A = Ba, Pb) and 0.8 $BiFeO_3$–0.2 $BaTiO_3$. J. Eur. Ceram. Soc. 24, 1551.
[18] J. S. Kim, C. I. Cheon, C. H. Lee and P. W. Jang. **2004.** Weak ferromagnetism in the ferroelectric $BiFeO_3$–$ReFeO_3$–$BaTiO_3$ solid solutions (Re = Dy, La), J. Appl. Phys. 96, 468.

[19] N. Gouitaa, T. Lamcharfi, L. Bouayad, F. Abdi, M. Ounacer and M. Sajieddine **2021**. Study of mutiphase structure and phases transition of $Ba_{0.95}Bi_{0.05}Ti_{1-x}Fe_xO_3$ ceramics at x=0.00 to 1.00 synthesis by solid state route. volume 18 and number 2. 10.22068/ijmse.18.2.3. (In press).
[20] Joby E. Joy, E. Atamanik, R. Mani, A. Nag, R.M. Tiwari, V. Thangadurai, J. Gopalakrishnan **2010**. $Ba_3M^{III}TiM^{V}O_9$ (M^{III} = Fe, Ga, Y, Lu; M^{V} = Nb, Ta, Sb) perovskite oxides: Synthesis, structure and dielectric properties. Solid State Volume 12, Issue 12, Pages 1970-1976.
[21] E. Mashkina, C. Mc Cammon, F. Seifert **2004**. A Mossbauer study of oxygen vacancy and cation distribution in 6H-$BaTi_{1-x}Fe_xO_{3-x/2}$. J. Solid State Chem. 177, 262.
[22] S. Ke, H. Huang, H. Fan, H.L.W. Chan, L.M. Zhou **2008**. Colossal dielectric response in barium iron niobate ceramics obtained by different precursors. Ceram. Int. 34, 1059.
[23] L.D. Landau and V.L. Guinzburg, Zh. Eksp.Teor **1950**. Fiz, 20, 1064.
[24] L.D. Landau and E.M. Lishfitz **1959**. Statistical Physics. Pergamon, London-Paris.
[25] Kun Li, Xiao Li Zhu, Xiao Qiang Liu, Xiang Ming Chen **2012**. Relaxor ferroelectric characteristics of $Ba_5LaTi_3Nb_7O_{30}$ tungsten bronze ceramics. Appl. Phys. Lett. 100 (1) 012902.
[26] L.E. Cross **1994**. Relaxorferroelectrics: An overview. Ferroelectrics 151, 305e320.
[27] N.S. Echatoui, T. Lamcharfi, S. Sayouri, L. Hajji, A. Alimoussa **2005**. Structural study of the $PbZr_{0.52}Ti_{0.48}O_3$ under Hydrostatic Pressure Phys. Chem. News, 26, 40-46.
[28] M. Villegas, A.C. Caballero, P. Duràn and F. Fernandez **2000**. Influence of Processing Parameters on the Sintering and Elictrical Properties of $Pb(Zn_{1/3}Nb_{2/3})O_3$-Based Ceramics. J. Am. Ceram. Soc. 83 (1) 141-66.
[29] K. Funke **1992**. Jump relaxation in solid electrolytes. Solid State Ionics 18–19, 183 1986; Prog. Solid State Chem. 22, 111.
[30] Z. Yu and C. Ang **2002**. Maxwell–Wagner polarization in ceramic composites $BaTiO_3$–$(Ni_{0.3}Zn_{0.7})Fe_{2.1}O_4$. Journal of Applied Physics, 91(2), 794–797.

Effect of dielectric screening function on binding energy and diamagnetic susceptibility of impurity doped in GaAs conical quantum dots.

E Iqraoun [1], A Sali [1] and K El – Bakkari [1]

[1] Department of Physics, Sidi Mohamed Ben Abdellah University, Faculty of Science, B.P.1796. Dhar El Mahraz, Fez, Morocco.

Corresponding author. elhassaniqra@gmail.com

Abstract. In this study, we have investigated the effect of the impurity dependent dielectric screening function on the diamagnetic susceptibility and the binding energy of a shallow-donor impurity confined in a GaAs conical quantum dot (CQD). The calculations have been made within the effective mass approximation, variational method and considering an infinite confining potential. Our results suggest that the effects of the variation of geometrical parameters of CQD are playing a pronounce influence on the diamagnetic susceptibility and the binding energy. The results show that the diamagnetic susceptibility depends strongly on the impurity position and it severely affected by the geometrical parameters of CQD.

Keywords: Diamagnetic susceptibility; binding energy; conical quantum dot; impurity dependent dielectric screening function.

1. Introduction

In the last decades, semiconductors of low-dimensional (nanostructures) stay one of the most intensely studied objects because of their peculiarity with the development of new techniques for nanofabrication like laser beam, Stranski Krastanov, chemical lithography and molecular beam epitaxy. [1-3]. Among these nanostructures we find, quantum dots (QDs) with various geometrical such as spherical, cylindrical, ellipsoidal, pyramid-like, lens-shaped and cone-like and other nanostructures which have received lots of attention due to their particular electronic and optical properties. These properties can be modified by different factors such as hydrostatic pressure, temperature, the position of impurity atoms, the geometry of the structure and the effects of static magnetic or electric fields, etc. This low-dimensional quantum dot has been used in a various modern technological application [4-8].

IOP Conf. Series: Materials Science and Engineering 1160 (2021) 012007 doi:10.1088/1757-899X/1160/1/012007

In low dimensional semiconductor systems, the spatial dependent screening of an impurity ion caused by the valence electrons and the effect of dielectric mismatch at the interfaces play a very important role [9-13]. A. John Peter et al. [14] calculated the effects of impurity position-dependent effective mass and dielectric function of a hydrogenic donor in a quantum dot. Their results indicated that the inclusion of the spatially varying mass and dielectric function results to an increment of the binding energy. S. Rajashabala et al. [15] have investigated the effects of dielectric screening and position dependent effective mass on the donor binding energy and the diamagnetic susceptibility in a quantum well. They found that the incorporation of the effect of the impurity position dependent effective mass on the donor binding energy is more significant for systems with thin well dimensions.

In the present work, we have investigated the effect of impurity dependent of dielectric screening function on the binding energy and diamagnetic susceptibility of a hydrogenic donor impurity located in a GaAs conical semiconductor quantum dot. The results have been presented as a function of the geometrical dimensions. For this purpose, the binding energy and corresponding wave functions of an electron confined in conical QDs have been calculated by employing a variational method. The paper is constituted as follows: Section 2 contains the theoretical framework. In Section 3, we present the results and their discussion. Finally, the conclusions are given in Section 4.

2. Theoretical framework

We consider an impurity donor confined in a GaAs conical quantum dot. The Hamiltonian describing this impurity is treated in the effective mass and the parabolic band approximations:

$$H_0 = -\frac{\hbar^2}{2m^*}\Delta + V_{conf}(x,y,z) - \frac{e^2}{\varepsilon(r)\,r}, \tag{1}$$

where m^* and $\boldsymbol{\varepsilon(r)}$ are the electron effective mass and impurity dependent dielectric screening function respectively. $V_{conf}(x,y,z)$ is the confining potential of the electron in CQDs which can be expressed by the following form :

$$V_{conf}(x,y,z) = \begin{cases} 0 \; inside \\ \infty \; outside \end{cases} \tag{2}$$

with $a = tan^{-1}(\phi/2)$ and ϕ the apical angle of the cone. In the absence of the impurity centre, we make a change to a new set of variables: $u = \rho\cos(\phi)$, $v = \rho\,sin(\phi)$, $\rho = \sqrt{u^2+v^2}$ and $w \in [0,h]$ where $\rho \in [0, 1/a]$. Then, Laplace operator in variables given with these transformations can be written in the form:

$$\Delta_{\phi\rho w} = \frac{1}{w^2}\left(\frac{\partial^2}{\partial\rho^2} + \left(\frac{1}{\rho} + \frac{\rho}{\rho^2+1}\right)\frac{\partial}{\partial\rho} + \frac{1}{\rho^2}\frac{\partial^2}{\partial\phi^2}\right) + \frac{1}{w^2(\rho^2+1)}\frac{\partial}{\partial w}\left(w^2\frac{\partial}{\partial w}\right), \tag{3}$$

where $\rho \in [0,1/a]$.

Owing to the cone symmetry, the following transformations will be used for $\phi < 90$. In this case, the parameter ρ will change in the interval. $0 \le \rho < 1$. The Equation (3) under the mentioned conditions will reduce to a set of three independent equations associated to each of the new coordinates $\Psi_\phi(\phi)$, $\Psi_\rho(\rho)$ and $\boldsymbol{\Psi}_\omega(\omega)$.

$$\Psi(\phi,\rho,\omega) = \Psi_\phi(\phi)\Psi_\rho(\rho)\Psi_\omega(\omega), \tag{4}$$

$$\begin{cases} \frac{\partial^2 \Psi_\phi}{\partial \phi^2} + m^2 \Psi_\phi = 0 \\ \frac{\partial^2 \Psi_\rho}{\partial \rho^2} + \frac{1}{\rho}\frac{\partial \Psi_\rho}{\partial \rho} + \left(k_\rho^2 - \frac{m^2}{\rho^2}\right) = 0 \\ \frac{1}{\omega^2}\frac{\partial \Psi_\omega}{\partial \omega}\left(\omega^2 \frac{\partial \Psi_\omega}{\partial \omega}\right) + \left(k^2 - \frac{k_\rho^2}{\omega^2}\right) = 0 \end{cases} \quad (5)$$

It is clear that the solution of the first equation of the system has an exponential form. $\Psi_\phi = C_1 e^{im\phi}$. The solutions of the second and third equations of the system can be presented via the Bessel function, namely, $\Psi_\rho = C_2 J_m(k_\rho \rho)$ and $\Psi_\omega = (C_3/\sqrt{\omega}) J_v(k\,\omega)$. Thus, the solution of the wave equation for the conical quantum dot in Cartesian coordinate system can be rewritten as [16]:

$$\Psi(x, y, z) = (C_c /\sqrt{z}) J_v(k\,z) J_m\left(k_\rho \frac{\sqrt{x^2 + y^2}}{z}\right) e^{im\phi}, \quad (6)$$

This wave function is the product of the exponential term and a hydrogenic-like wave function term. The wave function of the correlated electron–impurity can be written as:

$$\Psi_0(x, y, z) = N\Psi_1(x, y, z) exp(-\alpha r), \quad (7)$$

with $r = \sqrt{x^2 + y^2 + (z - z_i)^2}$,

$\boldsymbol{N}$ is the normalization constant and $\boldsymbol{\alpha}$ is a variational parameter.

2.1. The Effect of Dielectric Screening Function:

The Impurity position-dependent dielectric screening function is given by:

$$\frac{1}{\varepsilon(r)} = \frac{1}{\varepsilon_0} + \left(1 - \frac{1}{\varepsilon}\right) e^{-\frac{r}{\beta}}, \quad (8)$$

where $\boldsymbol{\varepsilon_0}$ is the static dielectric constant and $\boldsymbol{\beta}$ is the screening constant [17.18].

The Hamiltonian when the effect of dielectric screening function is included can be written as:

$$H = -\frac{\hbar^2}{2m^*}\Delta + V_{conf}(x, y, z) - \frac{e^2 \varepsilon_0}{r\,\varepsilon(r)}, \quad (9)$$

The diamagnetic susceptibility of the hydrogenic donor is given by:

$$\chi_{dia} = -\frac{e^2}{6m^* \varepsilon(r)\, c^2} < r^2 >, \quad (10)$$

where c is the velocity of light and $< r^2 >$ is the mean square distance of an electron from the nucleus [19,20].

IOP Conf. Series: Materials Science and Engineering 1160 (2021) 012007 doi:10.1088/1757-899X/1160/1/012007

3. Results and discussion:

We compute the donor impurity ground state binding energy and diamagnetic susceptibility in the GaAs conical quantum dot, with two fixed values of the apical angle ($\theta = 30°\,\&\,50°$) and for an impurity localized at $(0,0,2\,h/3)$. The input parameters for a prototypical GaAs CQD are $\varepsilon = 12.35$, $e = 1.6 \times 10^{-19}C,\ c = 3 \times 10^{8}m/s$ and $m^* = 0.0665m_0$ where m_0 is the free electron mass.

Figure 1 shows the calculation of the binding energy of the impurity, versus the height of the CQD, with two fixed values of the apical angle ($\theta = 30°\,\&\,50°$) and for two different cases with constant dielectric (full line), (dashed line) and with impurity dependent dielectric screening function (dotted line), (dashed-dot line).We see from the figure that the binding energy decreases as the height of CQD increases in all cases. This can be attributed to the weakening of the coupling electron–impurity with the increase of the height. Additionally, the decreasing of the apical angle starts making the electron in a smaller space (the increase of confinement), as a result, it increases the binding energy of impurity, especially for smaller cone. Moreover, it can be observed from the figure that the dielectric screening function affect strongly the impurity binding energy. Indeed, the incorporation of dielectric screening function diminishes the impurity binding energy particularly for a small cone, and this effect enhances as the apical angle of CQD becomes smaller. This behavior is the same as in the case of impurity confined in a quantum well [21] and impurity in a spherical QD [14].

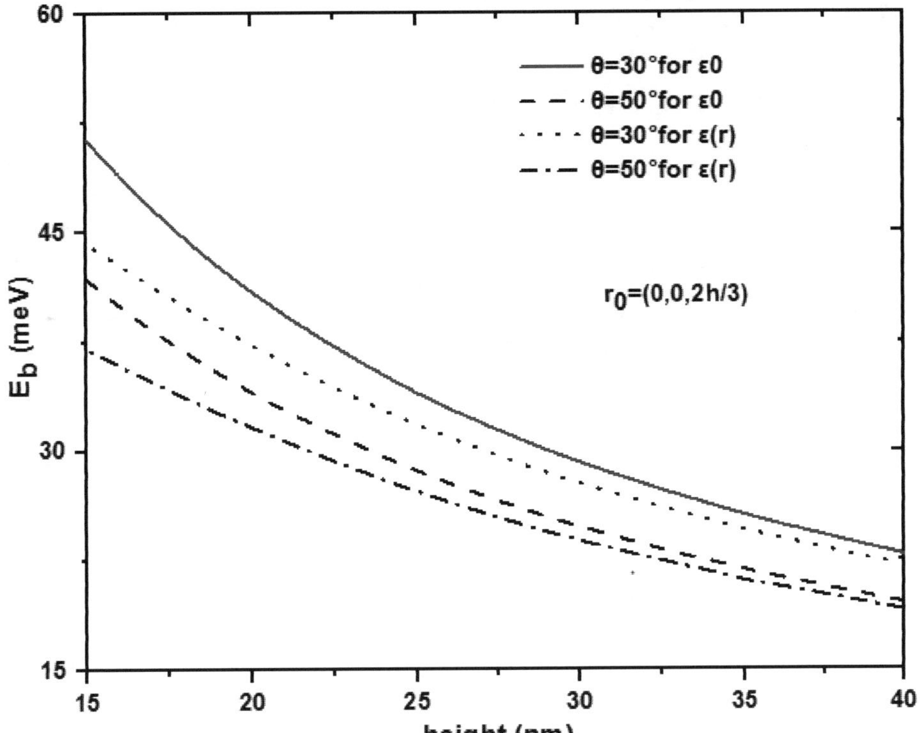

Figure 1. Variation of the binding energy of a donor impurity in a *GaAs* CQD as a function of the height (h) of the cone with constant dielectric (full line), (dashed line) and with impurity dependent dielectric screening function (dotted line), (dashed-dot line) for two fixed values of the apical angle ($\theta = 30°\ \&\ 50°$) and for an impurity position $(0,0,2\,h/3)$.

In figure 2, we present the variation of the difference in the binding energy versus the QD height of an impurity confined in a CQD with and without the including the impurity dependent dielectric screening function as a function of the height of the quantum dot for two fixed values of the apical angle ($\theta = 30° \,\&\, 50°$). It is clear from this figure that the dielectric screening function decreases the binding energy especially when the height of the structure becomes smaller, we can explain this behavior by the fact that the Coulomb interaction between the electron and the impurity diminishes due to the gradual weak localization of the electron wave function in the CQD, which leads to a decline of the impurity binding energy [22].

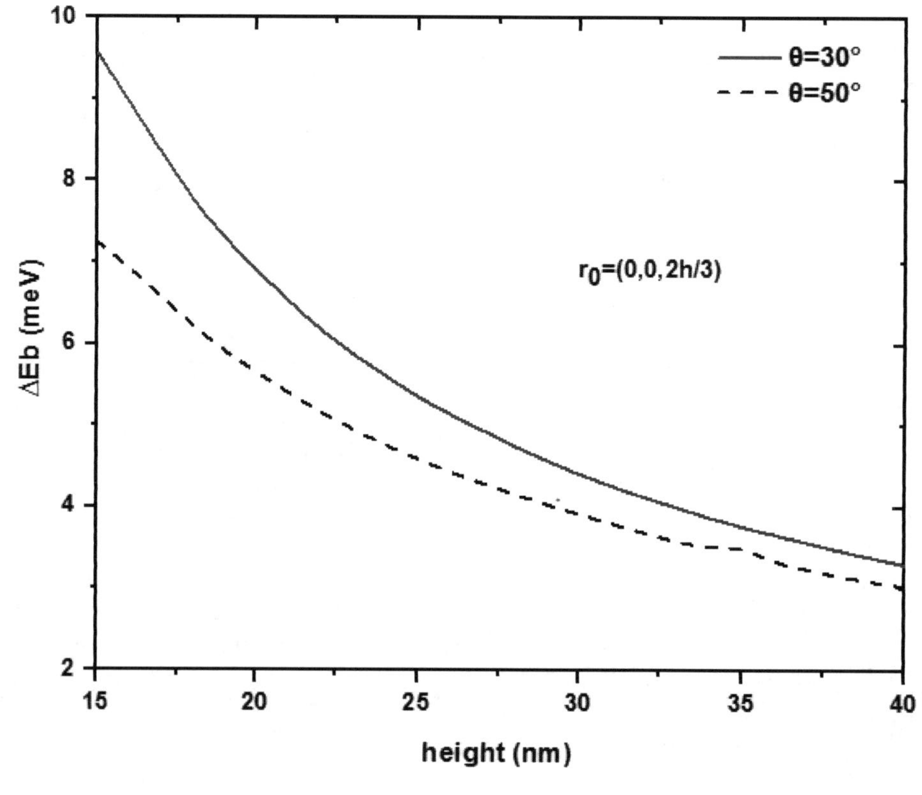

Figure 2. The variation of the difference in the binding energy with excluding and including impurity dependent dielectric screening function as a function of the cone height for two fixed values of the apical angle ($\theta = 30° \,\&\, 50°$).

In figure 3, we depict the results of the magnitude $< r^2 >$ of the diamagnetic susceptibility χ_{dia} as a function of the cone height of an impurity donor confined in a $GaAs$ CQD with constant dielectric (full line), (dashed line) and with impurity dependent dielectric screening function (dotted line), (dashed-dot line) with the same parameters as in the figure 1. It is observed from the figure that for fixed value of the apical angle, the magnitude of the diamagnetic susceptibility with and without dielectric screening increases with increasing the dot size. We can explain this behavior by the reason that when the dimensions of the CQD augment, the wave function of impurity becomes less concentrated around the electron which leads to an increment on the magnitude of the diamagnetic susceptibility. The figure reveals also that for fixed values of the height and the apical angle, the expectation value of $\langle r^2 \rangle$ increases slightly with the inclusion of the dielectric screening function especially for small height. The latter behavior affirms the diminution of impurity binding energy with the dielectric screening function that observed in figure 1. These results agree well with those obtained for the case in low dimensional semiconducting systems [23].

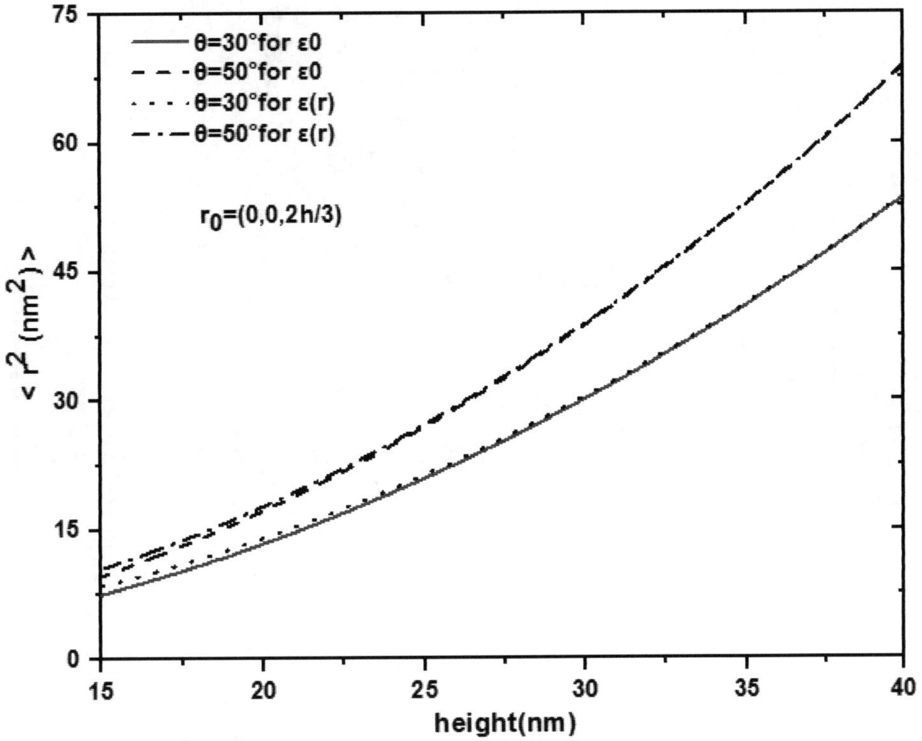

Figure 3. The variation of the magnitude of diamagnetic susceptibility as a function of the cone height with constant dielectric (full line), (dashed line) and with impurity dependent dielectric screening function (dotted line), (dashed-dot line) for an impurity position located at $(\mathbf{0}, \mathbf{0}, \mathbf{2}\boldsymbol{h}/\mathbf{3})$.

Figures 4 and 5 present the variation of the diamagnetic susceptibility and the difference between the diamagnetic susceptibility with constant dielectric (full line), (dashed line) and with impurity dependent dielectric screening function (dotted line), (dashed-dot line) respectively as a function of the height of the quantum dot, for two fixed values of the apical angle ($\theta = 30° \,\&\, 50°$) and for an impurity localized at $(0, 0, 2h/3)$. As it can be seen from figure 4, the diamagnetic susceptibility is a decreasing function of the cone height [23,25]. Furthermore, the same figure indicates that there is no appreciable change in the diamagnetic susceptibility for thick dots when the dielectric screening function is introduced. Additionally, it should be noted that the incorporation of the dielectric screening function on the diamagnetic susceptibility is important for smaller CQD (figure 5). These results in the presence of dielectric screening effect are in accordance with the previous investigation carried out by P. Nithiananthi and co-workers [23]. Their study has been focused on the effect of dielectric screening on the diamagnetic susceptibility of a donor in low dimensional semiconducting systems using the variational method within the effective mass approximation.

IOP Conf. Series: Materials Science and Engineering 1160 (2021) 012007 doi:10.1088/1757-899X/1160/1/012007

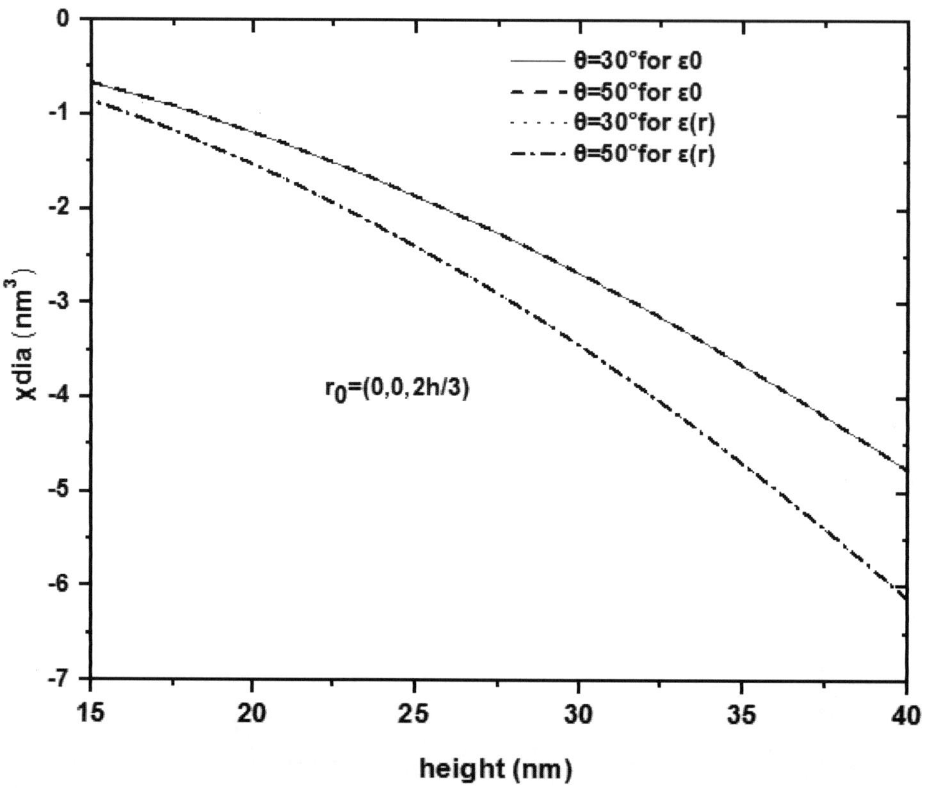

Figure 4. The diamagnetic susceptibility of an impurity with constant dielectric (full line), (dashed line) and with impurity dependent dielectric screening function (dotted line), (dashed-dot line) as a function of the cone height for two fixed values of the apical angle ($\theta = 30° \,\&\, 50°$) and for an impurity position $(0, 0, 2\,h/3)$.

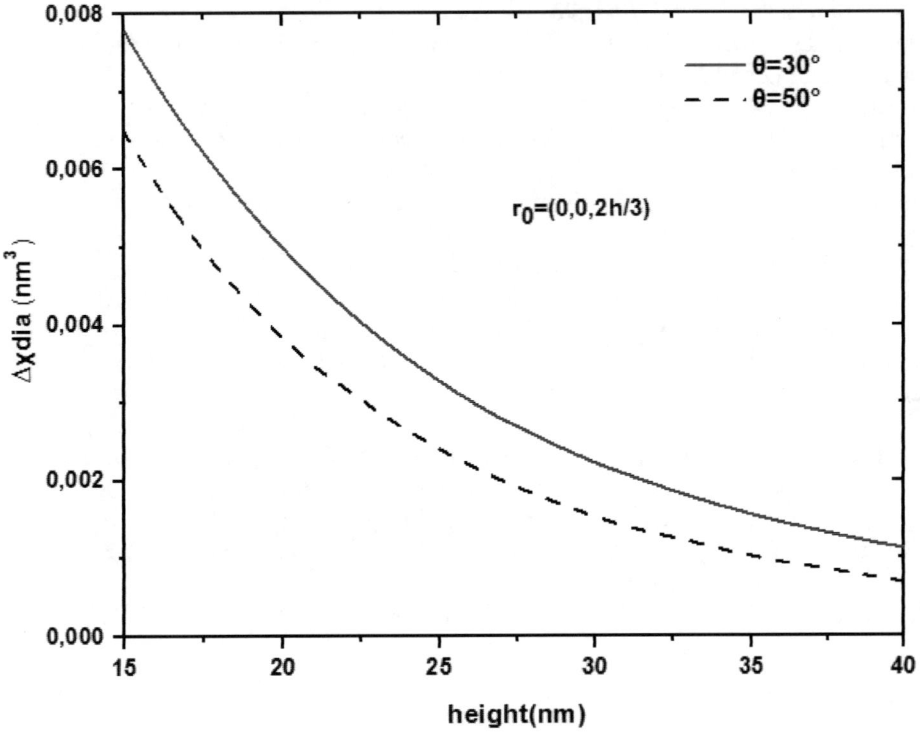

Figure 5. The variation of the difference between the diamagnetic susceptibility with including and excluding the impurity dependent dielectric screening function conversely to figure 1, as a function of the cone height for two fixed values of the apical angle ($\theta = 30° \,\&\, 50°$) for an impurity position $(0, 0{,}2\,h/3)$.

4. Conclusion

We have investigated the effect of dielectric screening function on the binding energy and the diamagnetic susceptibility of a donor impurity confined in a conical GaAs quantum dot as a function of the geometrical parameters, by utilizing the variational method within the effective-mass approximation. We have shown that the binding energy and the diamagnetic susceptibility are strongly dependent on the size of the dot. Additionally, we have found that the inclusion of the effect of dielectric screening function leads to a diminution of the donor binding energy and the diamagnetic susceptibility. Also, we have shown that the diamagnetic susceptibility decreases drastically for large values of the dimension of the dot. Moreover, the magnitude of the diamagnetic susceptibility increases with inclusion of the effect of dielectric screening function.

References

[1] Iqraoun E, Sali A, Rezzouk A, Feddi E, Dujardin F, Mora-Ramos ME & Duque CA 2017 *Philos. Magaz.* ***97*** 1445.
[2] Shi L, Yan Z.W 2016 *Superlatt. and Microstruc.* **94** 204.
[3] Cristea M, Radu A and Niculescu E. C 2013 *J. Lumin.* **143**, 592.
[4] El-Bakkari K, Sali A, Iqraoun E, Rezzouk A, Es-Sbai N and Ouazzani Jamil M 2018 *Physica B: Condensed Matter* **538** 2018.
[5] Chang-Hasnain C. J and Chuang S. L 2006 *J. Lightwave Technol.* **24** 4642.

[6] Ohgaki T et al. 2003 *J. Appl. Phys.***93**, 1961.
[7] Hayrapetyan D B, Kazaryan E M and Sarkisyan H A, 2016 *Opt. Communic.* **371**, 138.
[8] Nann Tand Skinner W M, 2011 *ACS Nano* **5**, 5291.
[9] John peter A, Navaneethakrishnan K, 2000 *Phys. Status Solidi B.* **220**, 897.
[10] Nithananthi P, Jayakumar K, 2006 *Solid State Commun.* **137**, 427.
[11] Oliveira L E, Falicov LM, 1986 *Phys. Rev. B.* ***34***, 8676.
[12] Rajashabala S and Navaneethakrishnan K, 2006 *Int. J. Mod. Phys. B.**20***, 1529.
[13] Rezaei G, Doostimotlagh N A,2012 *Physica E* **44**, 4.
[14] Peter AJ and Navaneethakrishnan K, 2008 *Physica E* .**40**, 2747.
[15] Rajashabala S and Navaneethakrishnan K, 2006 *Int. J. Mod. Phys. B.**20***, 1529.
[16] Lozovski Vand Piatnytsia V, 2011 *J. Comput. Theoret. Nanosc.* **8**, 11.
[17] Rejo Jeice A, Gerardin Jayamy Sr, Joph Wilsonz K.S, 2018 *Int. J. Mod. Phys. B* **32**, 1850122.
[18] Kilicarslan, E. et al. 2009 *Phys. E*, **42**.5.
[19] Akbas H, Aktas S, Okan SE and Ulas M, 1998 *Solid State Commun.* **23**, 113.
[20] El-Bakkari, K., Sali, A., Iqraoun, E. & Ezzarfi, 2020 *A. Superlattice Microst.* **148**, 106729.
[21] Abarna R, Anitha A and Arulmozhi M, 2017 *Journal of Physical Science*, Vol. 28(1).
[22] Deng ZY, Guo JK, Lai TR, 1994 *Phys. Rev. B*, **50**.8.
[23] Nithiananthi, P., Vickraman, P. & Jayakumar, 2009 *Int. J. Modern Phys. B*, **23**.8.
[24] Jayam G and Navaneethakrishnan K, 2003 *Solid State Commun.* **126**, 681.
[25] Latha M, Rajashabala S & Navaneethakrishnan, 2006 *K Phys. Stat. Sol. B*, **243**.6.

Optical absorption study of manganese-doped Na_2O-K_2O-P_2O_5 glasses

M Jerroudi[1], L Bih[2], M Haddad[3] and I Saadoune[4,5]

[1]Equipe de Physico-Chimie de la Matière Condensée, PCMC, Faculté des Sciences de Meknès, Maroc.
[2]Laboratoire des Sciences et Métiers de l'Ingénieur (LSMI), département matériaux et procédés, ENSAM, Université Moulay Ismail, Maroc.
[3]Laboratoire de Spectrométrie, des Matériaux et Archéomatériaux (LASMAR), Université Moulay Ismail, Maroc.
[4]Materials Science and Nano-engineering, Mohammed VI Polytechnic University, Lot 660 Hay Moulay Rachid, Ben Guerir, Morocco.
[5]IMED-Lab., Cadi Ayyad University (UCA), Av. A. El Khattabi, P.B. 549, Marrakesh, Morocco.

Corresponding e-mail address: meryjerroudi1994@gmail.com

Abstract.

The mixed alkali phosphate glasses with the general formula $49.95[xNa_2O\text{-}(1\text{-}x)K_2O]\text{-}0.1MnO_2\text{-}49.95P_2O_5$ with x=0-1mol% were prepared and characterized using the optical absorption technique. The optical absorption spectrum of each glass exhibits a single band in the region around 200-400 nm. From the ultraviolet absorption edges, the optical parameters including optical band gap and the Urbach energy were evaluated. It is interesting to note that these optical parameters are composition-dependent. The optical band gap energy (E_g) and the Urbach energy (ΔE) show the presence of a maximum and a minimum respectively around the intermediate composition x=0.5. The non-linearity behavior of the composition dependence of E_g and ΔE is a fingerprint of the mixed-alkali effect in the glasses under study.

Keywords: Manganese, Phosphate, glasses, Optical properties, MAE.

1. Introduction

Glass is the most frequently used material in several applications fields. Topical progression in the glass field facilitates the researcher to design numerous kinds of oxide glasses such as borate, phosphate, tellurite, silicate, germinate, etc. Due to the specific properties of phosphates glasses, such as low thermal characteristics, higher thermal expansion coefficients, UV transparent and they can incorporate large amounts of oxides [1,2], they are considered as the more important glasses compared to other glass families. So, the study of their properties is a crucial interest in order to use them in various fields of applications.

Most of the properties of the glassy materials are depending on the chemical composition of the glass system and the non-linear behaviour of certain properties with the substitution of one alkali by a dissimilar one while keeping the total alkali content constant is known as the mixed alkali effect [3]. The most evident manifestation of this effect has been observed in the variation of some dynamic properties. The experimental and theoretical investigations on the mixed alkali effect (MAE) of several glasses are reported in the literature [4,5]. The novelty of this work is double: firstly, to

formulate new low melting phosphate glasses; secondly, to understand the effect of the substitution of potassium by sodium on the optical properties of the glasses containing manganese oxide.
Recently [6], we have reported on the thermal and structural properties of the $49.95[xNa_2O\text{-}(1\text{-}x)K_2O]\text{-}0.1MnO_2\text{-}49.95P_2O_5$ with (x=0-1 mol%) glasses by Differential Scanning Calorimetry (DSC), Fourier-Transform Infrared (FTIR), Raman, and Electron Paramagnetic Resonance (EPR). The above technique highlights the presence of the mixed alkali effect in these glasses. The foremost purpose of this work is to study the effect of mixed Na_2O and K_2O oxides on the optical properties of the above manganese phosphate glasses.

2. Experimental procedure

The glasses compositions in the system $49.95[xNa_2O\text{-}(1\text{-}x)K_2O]\text{-}0.1MnO_2\text{-}49.95P_2O_5$ with x=0-1mol% (labelled $Na_xK_{1-x}PMn$) are prepared using the conventional melt quenching technique. The chemical reagent precursors are di-hydrogen ammonium phosphate ($NH_4H_2PO_4$) (Aldrich, 99%), potassium carbonate (K_2CO_3) (LobaChemie, 99%), sodium carbonate (Na_2CO_3) (LobaChemie, 99%), and manganese oxide (MnO_2) (Aldrich, 99.9%). The appropriate amount of the different reagents is crushed and grinded to fine powders. Details of the procedure adopted in the elaboration of these glassy-samples have been described elsewhere [6]. The obtained glasses are homogeneous transparent and colorless.
For the study of the optical properties of the glasses, a UV–Visible spectroscopy was performed on a Jascov-570 spectrophotometer over the spectral range of 200-800 nm. A barium sulfate ($BaSO_4$) plate was used as the standard on which the finely ground sample from the glass was coated.

3. Results and discussions

3.1. Absorption spectra of the glasses

The absorption spectra of the studied glasses $Na_xK_{1-x}PMn$ were recorded at room temperature in the region 200-800 nm and are displayed in Figure 1. These spectra have the same shape. They are characterized by a single absorption band over the region of 200–400 nm. As the amount of sodium oxide increases in the glassy matrix, one can note a shift of this absorption band to lower wavelengths. This shift may result from the variation of the content of the non-bridging oxygen atoms in the glassy system as sodium oxide increases. The present absorption band can be assigned to the presence of trace unavoidable of iron impurities in the raw materials and specifically to the predominance of trivalent Fe^{3+} ions [7]. Also, it may be due to the transfer charge (TC) between the oxygen 2p and the 3d orbital of manganese ion [8].
Furthermore, it is worth noting that no visible band could be detected in the region 400-800 nm in the optical spectra of the investigated glasses (Figure 1). This shows that the studied glasses are transparent in this region. This transparency is concomitant with the colorless appearance of these materials. However, the absence of color by visualization, although the glasses contain manganese, is due to the low content of the Mn^{2+} ion (0.1 mol %).

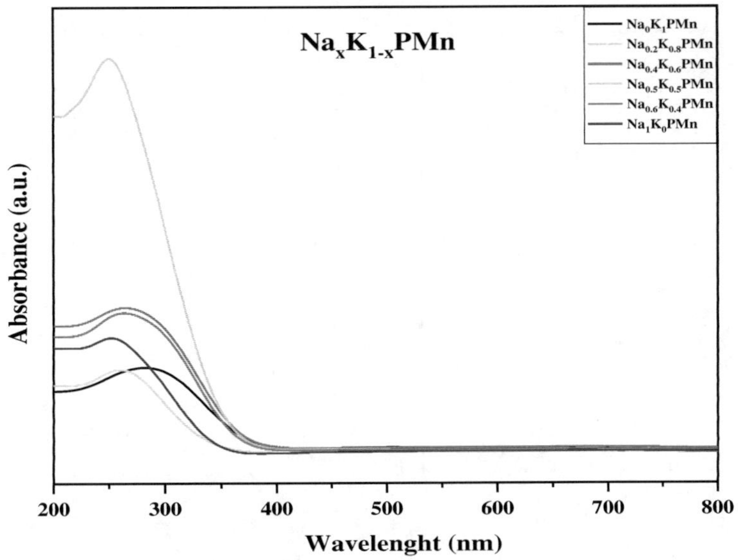

Figure 1. Optical absorption spectra in the range 200–800 nm for the $Na_xK_{1-x}PMn$ glasses.

3.2. Optical parameters of the glasses

From the absorption ultraviolet edges, the optical parameters including optical band gap energy and the Urbach energy were determined for the studied glasses.
The optical band gap energy values for the present samples were determined by employing the Tauc's plots method. The absorption coefficient $\alpha(v)$ near the edge is calculated according to the relation [9]:

$$\alpha\,(\upsilon)=(2.303/d)\ A(\upsilon) \qquad \textbf{(1)}$$

where "A" is the absorbance and "d" is the thickness of the sample (in centimeters).
Generally, for the amorphous and vitreous materials, Davis and Mott [9] described the optical absorption $\alpha(\upsilon)$ above the exponential tail by the following power-law expression:

$$\alpha h\upsilon = \alpha(h\upsilon - E_g)^n \qquad \textbf{(2)}$$

where "n" is an index that can have different values: 2, 3, 1/2, and 1/3 corresponding to indirect allowed, indirect forbidden, direct allowed, and direct forbidden transitions. "B" is a constant called a band-tailing parameter, "E_g" is the optical band gap energy and hν is the incident photon energy.
The Tauc's plots of the studied glasses $Na_xK_{1-x}PMn$ are represented in figure 2. The optical band gap energy values are determined by extrapolation of the linear part of the curves to the x-axis at zero absorption.
In order to estimate the degree of the disorder induced by the substitution of K_2O by Na_2O in the $Na_xK_{1-x}PMn$ glasses, we have explored the region of the lower photon energy of the edge for the glasses (Figure 3). In this region, the absorption coefficient changes into an exponential dependence, as given by the Urbach law [10]. The absorption coefficient α(ν) depends exponentially on the photon energy (hν). The exponential dependence known as the Urbach rule [10] may be written in the form:

$$\alpha\,(\upsilon) = B\,\exp\,(h\upsilon/\Delta E) \qquad \textbf{(3)}$$

Where ΔE is the Urbach energy and B is a constant.

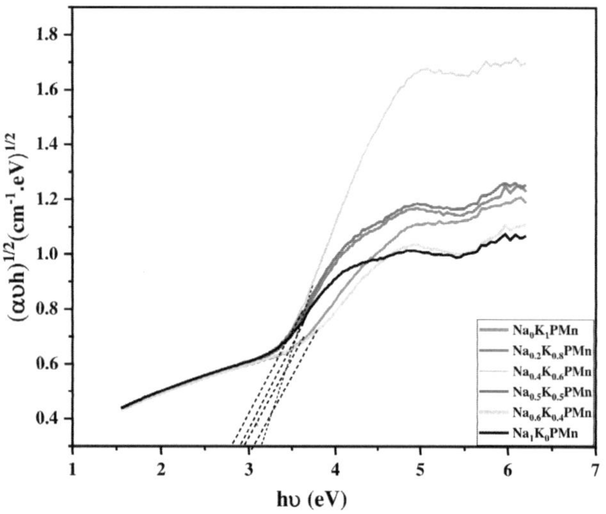

Figure 2. Tauc's plots for the $Na_xK_{1-x}PMn$ glasses.

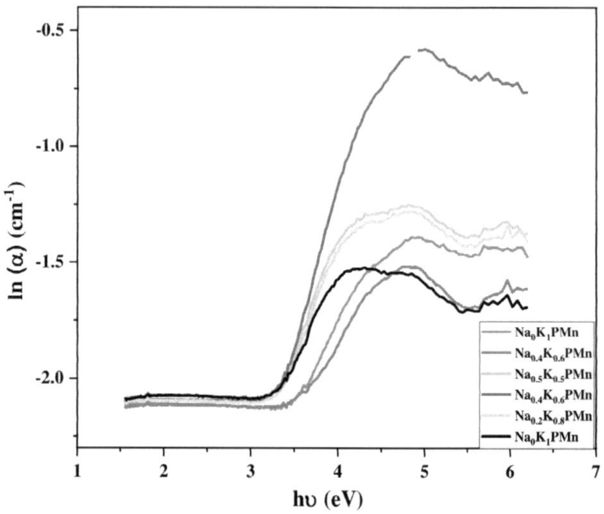

Figure 3. Variation of ln(α) as a function of photon energy (hυ) for the $Na_xK_{1-x}PMn$ glasses.

Figure 4 depicts the variation of the optical band gap energy and the Urbach energy as a function of Na/Na+K ratio. It is observed that the optical band gap energy E_g varies from 2.99 eV to 2.94 eV when K_2O is replaced by Na_2O. E_g presents a non-linear variation with the composition. It goes through a maximum at the intermediate composition x=0.5 ($Na_{0.5}K_{0.5}PMn$) (Figure 4). Namely, E_g goes through a maximum when the content of Na_2O is equal to K_2O in the studied glasses. The observed maximum of the E_g in the title phosphate glasses could be due to a distribution of defects resulting from the two alkali ions. A non-linear variation in the composition dependence of the optical band gap energy (E_g) has been observed with a maximum value at x=0.5 ($Na_{0.5}K_{0.5}PMn$). This non-linear variation indicates the existence of mixed alkali effect in the present glass system. This MAE could be explained by the formation of some defects in the band gap and the composition dependence of these defects varies none-linearily according to the K_2O/Na_2O ratio.

The variation of the absorption coefficient ln(α), as a function of energy hν, allows one to determine the Urbach energy for each glass. It is the inverse of the slope relative to the linear of the curve. The ΔE values are found to vary from 0.34 to 0.31 eV. The variation of the ΔE versus the composition for the $Na_xK_{1-x}PMn$ glasses reveals the presence of a minimum for the $Na_{0.5}K_{0.5}PMn$ glass. This discontinuity trend is an indicative to the presence of mixed alkali effect. It is interesting to note that the $Na_{0.5}K_{0.5}PMn$ glass exhibits the lower value of the Urbach energy, so one can conclude that the disorder is important for this glass.

It is evidenced that the composition dependence of the optical band gap energy and the Urbach energy are particularly sensitive to the substitution of potassium by sodium oxide and indicating the presence of mixed alkali effect in the studied glasses $Na_xK_{1-x}PMn$. Similar behaviour is also observed for other mixed alkali glasses [11,12].

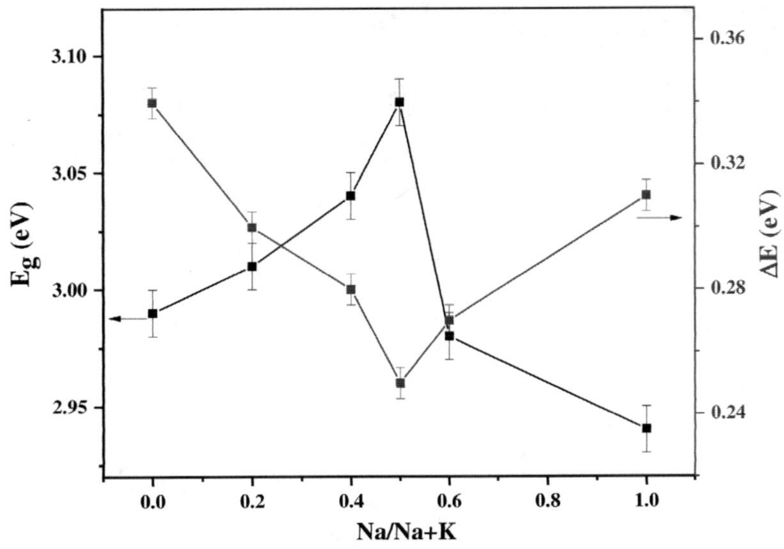

Figure 4. Variation of the band gap energy and Urbach energy with Na/Na+K ratio.

4. Conclusion

The optical properties of glasses belonging to the general formula 49.95[xNa_2O-(1-x)K_2O]-0.1MnO_2-49.95P_2O_5 with x= 0-1 mol% have been studied. The experimental results concerning absorption spectra of manganese-doped alkali phosphate glasses, show the presence of a single band in the region 200-400nm which is attributed to the iron impurities (Fe^{3+}) present in the vitreous composition and/or to the transfer charge (TC) between the oxygen 2p and the 3d orbital of manganese ion. The optical properties of the glass compositions showed that the optical band gap energy and the Urbach energy present maxima and minima, respectively when mixing equal amounts of K_2O and Na_2O in the Na_xK_{1-x}PMn. One can conclude that the optical properties are sensitive to the mixed alkali effect in the glasses under study.

Acknowledgments

The authors would like to thank the CNRST (Morocco) and OCP foundation for their financial support of this work in the framework of PPR (Energyphos) & Apphos (Mat - BIH-01/17) projects.

References

[1] Rajbhandari P, Montagne L, and Tricot G, (2015), *Mater. Chem. Phys.*, **155**, pp. 17–22.
[2] Jerroudi M, Bih L, Azrour M, Manoun B, Saadoune I, and Lazor P, (2019), *J. Inorg. Organomet. Polym. Mater.*
[3] Krishna Kumari G, Rama Krishna C, Muntaz Begum S, Pushpa Manjari V et al, (2013), *Spectrochim. Acta - Part A Mol. Biomol. Spectrosc.*, **101**, pp. 140–147.
[4] Abbas L, Bih L, Nadiri L, El Amraoui L, Mezzane D, and Elouadi B, (2008), *J. Mol. Struct.*, **876**, no. 1–3, pp. 194–198.
[5] Bih H, Es-soufi H, Bih L, and Bejjit L, (2019), *J of App Surf and Interf*, **4**, pp. 140-147.
[6] Jerroudi M, Bih L, Haily E *et al*, (2020), *Mater. Today Proc* **30**, pp. 1052-1055.
[7] Bae B and Weinberg M C, (1993), **7760**, pp. 1–7.
[8] Jerroudi M, Yousfi S, Bih L, Bejjit L, Haddad M, (2020), *IOP Conf.Series. MSE*, **783,** pp. 1-6.
[9] Taylor P, Davis E A, and Mott N F, (2006), *Philos. Mag.*, pp. 37–41.

[10] Urbach F, (1953), *Phys. Rev.* **92**, no. 3, p. 61.
[11] Chakradhar R P S, Yasoda B, Rao J L, and Gopal N O, (2006), *J. Non. Cryst. Solids*, **352**, no. 36–37, pp. 3864–3871.
[12] Srinivas G, Ramesh B, Siva Kumar J, Shareefuddin M, Chary M, and Sayanna R, (2016), *J. Taibah Univ. Sci.*, **10**, no. 3, pp. 442–449.

IOP Conf. Series: Materials Science and Engineering 1160 (2021) 012009 doi:10.1088/1757-899X/1160/1/012009

Structural properties of oil/water microemulsion with surface grafted PEO polymers

M Khatouri[1], R Ahfir[1], S Elkhaoui[1], M Lemaalem[2], L Talha[1], M Filali[1*]

[1]Laboratoire de Physique Appliquée, Informatique et Statistique (LPAIS), Faculty of Sciences Dhar El Mahraz, Sidi Mohamed Ben Abdellah University, BP 1796 Fes, Atlas, Morocco
[2]Laboratoire de Physique des Polymères et Phénomènes Critiques Sciences Faculty Ben M'Sik, Hassan II University, P. O. Box 7955, Casablanca, Morocco

*E-mail: mohammed.filali@usmba.ac.ma

Abstract: In this study, we investigated the structural properties of oil/water microemulsions with surface grafted polymers. In particular, we determined the effect of the addition of two kinds of hydrophobically modified polymers, PEO-m with one hydrophobic end and PEO-2m with two hydrophobic extremities. We used Molecular Dynamics simulations with the appropriate interaction potential that considers the VDW interactions, the electrostatic interactions, and the interaction induced by the polymers. Our results showed that the PEO-m introduces a steric repulsive interaction between the microemulsions in the diluted and concentrated systems, respectively, Φ=2.8% and Φ=14%. Thus, PEO-m favorises the stability and dispersion of the microemulsions. Besides, the PEO-2m introduces an attractive interaction between the droplets in the diluted case. Both PEO-2m and PEO-m strengthen the repulsive potential between the microemulsions in the concentrated case.

1. Introduction

Microemulsions are great candidates as potential drug delivery systems because they improve drug solubilization, long shelf-life, and ease of preparation and administration [1]. A microemulsion is defined as a thermodynamically stable liquid dispersion consisting of an aqueous and an oily phase combined with a surfactant [2,3]. It comes in the form of water droplets in oil or oil droplets in water, but these must be stabilized to prevent them from recombining because when two droplets come into contact, they combine naturally to minimize surface energy. To stabilize the emulsion droplets, amphiphilic molecules, such as surfactants, are often added, which reduce the interfacial tension in both phases. For drug delivery applications, nano-transporter are often covered with biocompatible polymers [4,5]. In this study, PEO is used in two different modified forms that we note, respectively, PEO-m and PEO-2m. The PEO-m refers to the PEO_{227} plus $C_{12}H_{25}$ alkane chain attached to one of its extremities. The latter permits graft of the PEO_{227} to the microemulsions. The PEO-2m denote the PEO_{227} plus two $C_{12}H_{25}$ grafted to each one of its extremities.

In the Molecular Dynamics simulation (MD), microemulsions are considered to be non-overlapping, uniformly charged nanospheres interacting via an effective potential based on the DLVO model, which is the sum of the repulsive potential of a hard-sphere, the van der Waals attractive term [6] and the Yukawa type screened Coulombic repulsive potential [7]. Besides, the presence of PEO-m with different quantities induces a steric-type repulsive interaction. The PEO-2m produces an additional attractive interaction potential related to the microemulsions bridging by the hydrophobic ends of the PEO-2m chains. The interaction potential parameters are already validated by the equation (OZ) method, with the approximation HNC [8].

In this work, we use molecular dynamics simulation to study the effect of the addition of the polymers PEO-m and PEO-2m on the structural and interaction properties of a spherical nanodroplet of positively charged O/W microemulsion. In this context, we use an ionic microemulsion comprising cetyl pyridinium chloride (CpCl) as a cationic surfactant, n-Octanol as co-surfactant, decane oil, and saltwater. We add a telechelic polymer PEO-m or PEO-2m to the microemulsion. This polymer consists of a water-soluble poly(ethylene oxide) (PEO), which has one or two hydrophobic ends ($C_{12}H_{25}$). PEO-m decorates the microemulsion nanodroplets, whereas PEO-2m has two possible configurations: bridging two different nanodroplets or decorating them.

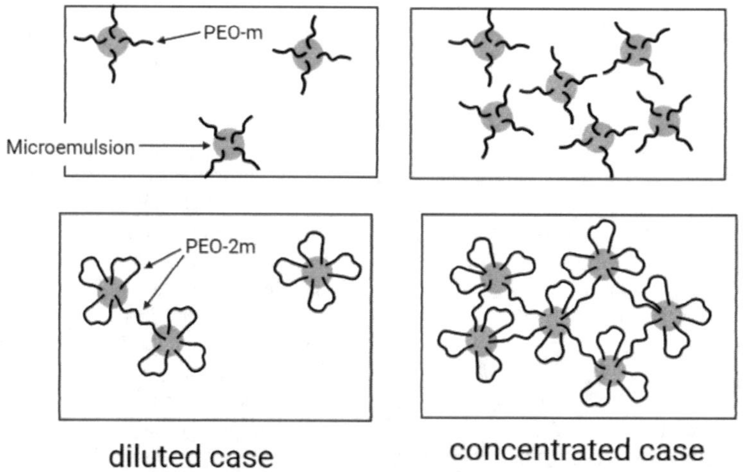

Figure 1. Schematic representation of two suspensions in the diluted and concentrated cases: Microemulsion droplets + polymer with one functional end (PEO-m); Microemulsion droplets + polymer with two functional ends (PEO-2m)

2. Molecular Dynamics simulation (MD)

2.1 Interaction potential

Molecular dynamics consists of calculating the evolution of an interacting particle system over time. This simulation is used as structural and dynamic models for understanding the experimental results, provided that the potential for interaction between the particles, which make up the system, is known.

In the case of a system consisting of ionic microemulsion, particles are modeled as uniformly charged hard spheres. The potential that models the interactions between each pair of microemulsions is the sum of a hard-sphere potential and a Van der Waals potential, and a Coulombic potential [9-10].

$$U(r) = \mathrm{U_{HS}} + \mathrm{U_{VDW}} + \mathrm{U_{coulomb}}$$

With U_{HS}: is the repulsive potential of a hard-sphere.

$$U_{HS} = \begin{cases} +\infty & , \quad r \leq 2R \\ 0 & , \quad r \geq 2R \end{cases}$$

U_{VDW}: is the attractive potential of Van der Waals

$$U_{VDW} = \begin{cases} \frac{-A_H}{6}\left[\frac{2R^2}{r^2-4R^2} + \frac{2R^2}{r^2} + \ln\left(\frac{r^2-4R^2}{r^2}\right)\right] & , \quad r \leq R_c \\ 0 & , \quad r > R_c \end{cases}$$

$U_{coulomb}$: A Yukawa-style coulombic repulsive potential:

$$U_{coulomb} = \begin{cases} K_B T \frac{z_{eff}^2 l_B}{1+kR} \frac{e^{-k(r-2R)}}{r}, & r \leq R_c \\ 0 & , \quad r > R_c \end{cases}$$

Where, r is the distance from centre to centre, $A_H = 1.1K_BT$ is the Hamaker constant, R=62Å is the radius of the microemulsions, Z_{eff} is the effective charge per nanodroplet and k^{-1} is the Debye length, $l_B = \frac{e^2}{4\pi\varepsilon_0\varepsilon_r K_B T}$ is the Bjerrum length, $\varepsilon_r = 80$ the dielectric constant of water and ε_0 with Permittivity of Vacuum, the temperature T will be set to the value of the ambient temperature 298 K and the effective charge per nanodroplet at $Z_{eff} = 130$, this value is taken from a previous structural study on this system revealed that an excellent quantitative agreement is found for all spectra with the same effective charge on each nanodroplet [11]. R_c is a cut off distance, of the interaction potential.

When PEO-m is added to the bare microemulsion, we have seen an additional repulsive contribution linked to the decoration of the nanodroplets by the PEO-m chains:

$$U_2 = U_{HS} + U_{VDW} + U_{coulomb} + V(r)$$

$V(r)$: A steric-type steric repulsive interaction of the Yukawa type:

$$V(r) = \begin{cases} V_{steric}\left(\frac{2R}{r}\right)e^{-\frac{r-2R}{\lambda_{steric}}}, & r \leq R_c \\ 0 & , \quad r > R_c \end{cases}$$

V_{steric} is the contact potential, λ_{steric} is the range of interaction we assume of the order of magnitude of the PEO radius of gyration.

And when we add the PEO-2m with two ends on the bare microemulsion. In this case, we have seen an additional attractive contribution linked to the bridging of the nanodroplets by the PEO-2m chains:

$$U_3 = U_{HS} + U_{VDW} + U_{coulomb} + V_{steric} + W_{bridging}$$

$W(r)$ Attractive Yukawa bridging type interaction:

$$W(r) = \begin{cases} W_{bridging}\left(\frac{2R}{r}\right)e^{-\frac{r-2R}{\lambda_{pontage}}}, & r \leq R_c \\ 0 & , \quad r > R_c \end{cases}$$

$W_{bridging}$ the contact potential and $\lambda_{pontage}$ the range of the interaction, which should be of the order of magnitude of the average distance between the two ends of the PEO chain (~100 Å) [16].

Figure 2 illustrates the interaction potential, which is used in MD simulations, schematised using the parameters discussed above, with ($V_{steric} = 2.6K_BT$ for Np=4(PEO-m), $V_{steric} = 5.2K_BT$ for Np=8(PEO-m)) and ($V_{steric} = 2.6K_BT$, $W_{bridging} = -0.53K_BT$ for Np=4(PEO-2m) and $V_{steric} = 5.2K_BT$, $W_{bridging} = -1.05K_BT$ pour Np=8(PEO-2m)) and σ = 2R.

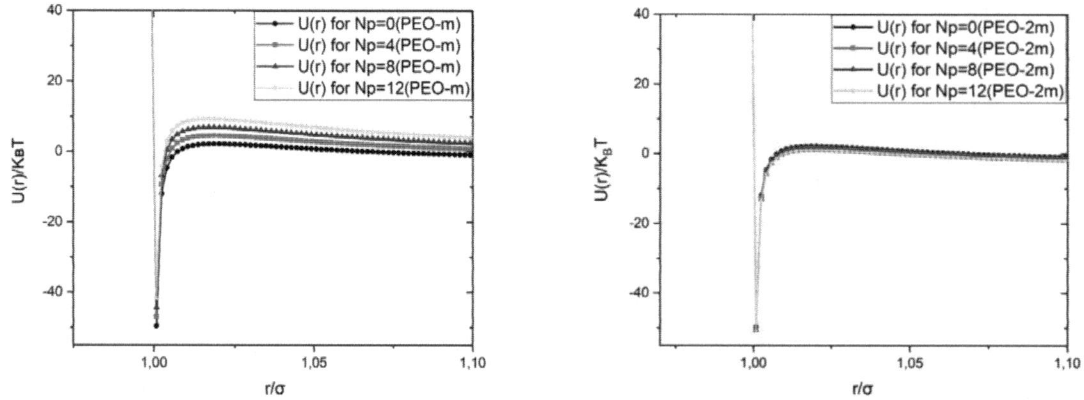

Figure 2. Dimensionless pair-potential between microemulsions, U(r)= K_BT, versus the dimensionless distance *r/σ* (σ = 2R).

In this work, NVT MD simulations were performed, using the LAMMPS simulation package [17], in a cubic box of volume V. We considered N = 10^6 microemulsions in the simulation box. We applied three-dimensional periodic boundary conditions to suppress the bord effect and simulate an infinite system [18].

3. Results and discussions

3.1 *Static properties*

Molecular dynamics is applied here to determine the structural properties of the bare microemulsion (Np=0), microemulsion with PEO-m polymer addition, and microemulsion with PEO-2m polymer addition.

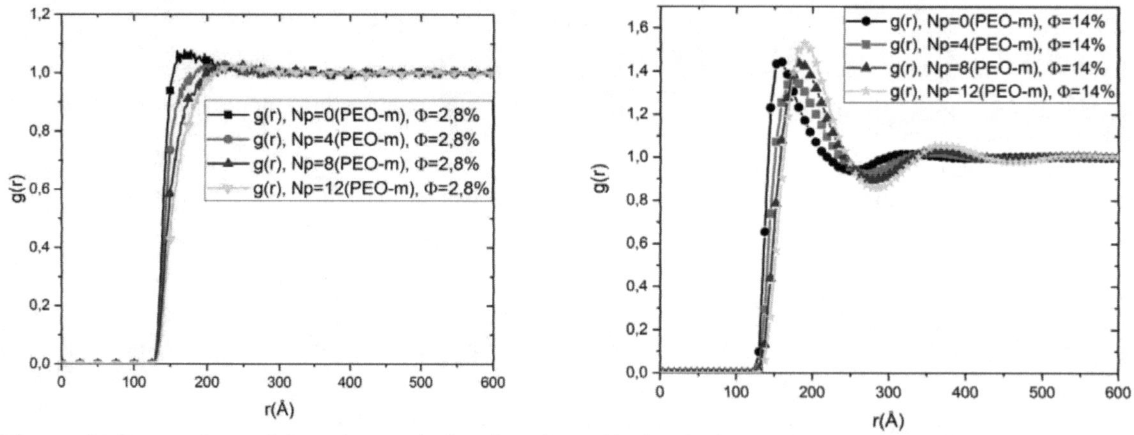

Figure 3. Comparison of the pair correlation function g(r), for the bare microemulsion (Np = 0) and microemulsions with added PEO-m (Np= 4, Np= 8, Np=12), for two volume fractions (Φ=2.8% and Φ=14%).

We present in figure 3 the radial distribution function g(r) for a mixed system of microemulsion and polymer PEO-m in the diluted case Φ=2.8% and the concentrated case Φ=14%. For Φ=2.8%, we observe a single correlation peak less intense and larger, which representing the interactions between the first close neighbors. We notice that when the parameter Np increases from 0 to 12(PEO-m), the correlation peak shifts towards greater distances, then the peak becomes less intense and larger. It varies from (gmax=1,06074, d ~ 165,54Å) to (gmax=1,02415, d ~ 230,02Å) see Table 1. For Φ=14%, we notice that the correlation function represents a more intense and narrower main peak representing the interactions between close neighbors and a second peak, less intense and wider, representing the interactions between second neighbors. The same remarks have been noticed when the parameter Np increases from 0 to 12(PEO-m), the correlation peak shifts towards greater distances, becomes more pronounced, and varies from (gmax=1, 44173, d ~ 159.96Å) to (gmax=1, 52909, d ~ 189,72Å). This result shows that the addition of PEO-m to the microemulsion introduces a repulsive contribution to the existing interaction through a certain interpenetration resistance between the microemulsions in the diluted case and the concentrated case, as can be seen from Figure 4.

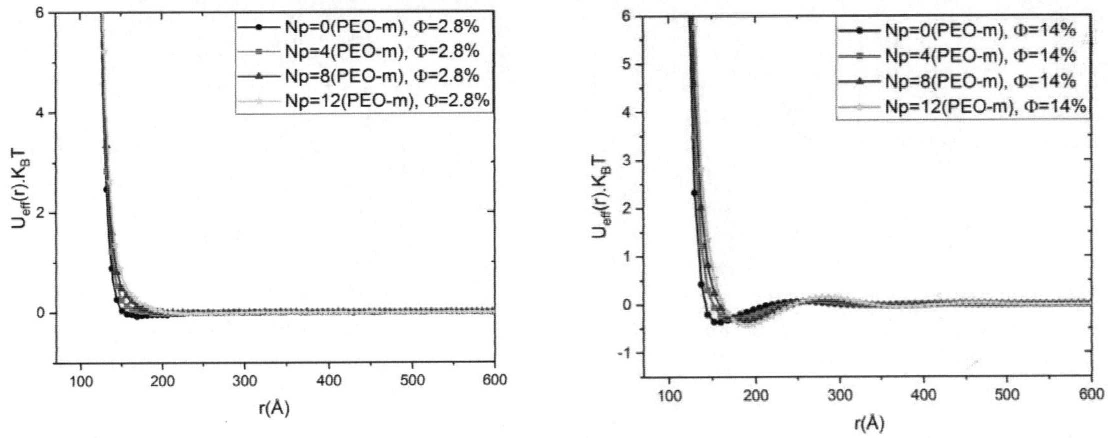

Figure 4. The mean field interaction potential $U_{eff}(r)=-k_BT\log(g(r))$ of microemulsions without PEO-m (Np=0) and with PEO-m (Np=4, 8 and 12) for Φ=2.8% and Φ=14%.

Table 1. Analysis of the g(r) curves obtained by MD for Np=0(PEO-m), 4(PEO-m), 8(PEO-m) and 12(PEO-m) in the diluted case (Φ =2.8%) and the concentrated case (Φ =14%).

	Φ=2.8%		Φ=14%	
	r_{max}	g_{max}	r_{max}	g_{max}
Np=0(PEO-m)	165.54 Å	1.06074	159.96 Å	1.44173
Np=4(PEO-m)	205.22 Å	1.03123	174.84 Å	1.35771
Np=8(PEO-m)	222 Å	1.02132	182.28 Å	1.43607
Np=12(PEO-m)	230.02 Å	1.02415	189.72 Å	1.52909

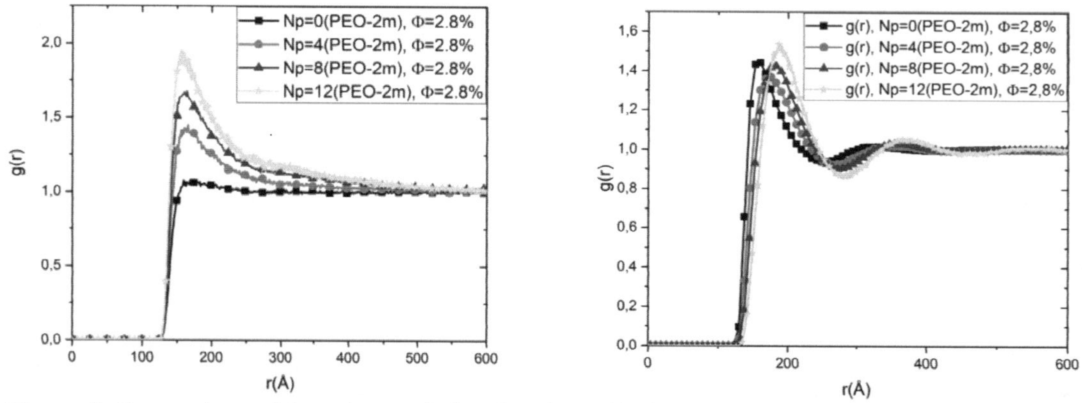

Figure 5. Comparison of the pair correlation function g(r), for the bare microemulsion (Np= 0) and microemulsions with added PEO-2m (Np = 4, Np = 8, Np=12), for two volume fractions (Φ=2.8% and Φ=14%).

In this part, we will discuss the effect of the POE-2m on the correlation function g(r) of the bare microemulsion for the diluted case Φ=2.8% and the concentrated case Φ=14%. In figure 5, we present the correlation function of a microemulsion Np=0(PEO-2m) and the correlation function of a microemulsion with polymer 4(PEO-2m), 8(PEO-2m), and 12(PEO-2m) in the diluted case Φ=2.8% and the concentrated case Φ=14%. For Φ=2.8%, we notice that the correlation peak shifts to smaller distances and becomes more intense and narrower as the Np parameter increases from 0(PEO-2m) to 12(PEO-2m) (see Table 2), which means the appearance of an attractive effective interaction between the microemulsions introduced by the POE-2m, as can be seen from Figure. 6 (left). But for the concentrated case Φ = 14%, the first peak becomes pronounced with the addition of the PEO-2m, and it moves to large distances (varies from 159.96Å to 186.55Å) when the Np parameter increases from 0(PEO-2m) to 8(PEO-2m). This shows that the addition of the PEO-2m to the bare microemulsion in the concentrated case introduces a repulsive contribution between the microemulsions, as can be observed in Figure 6 (right).

Table 2. Analysis of the g(r) curves obtained by MD for Np=0(PEO-2m), 4(PEO-2m), 8(PEO-2m) and 12(PEO-2m) in the diluted case (Φ =2.8%) and the concentrated case (Φ =14%).

	Φ=2.8%		Φ=14%	
	r_{max}	g_{max}	r_{max}	g_{max}
Np=0(PEO-2m)	165.54 Å	1.06074	159.96 Å	1.44173
Np=4(PEO-2m)	161.82 Å	1.6182	174.84 Å	1.36769
Np=8(PEO-2m)	158.1 Å	1.69437	182.28 Å	1.426
Np=12(PEO-2m)	154.53 Å	1.91	186.55 Å	1.515

These results showed that the addition of a PEO-m polymer on a bare microemulsion introduces a steric repulsive interaction between the nanodroplets, in the case of a concentrated system Φ=14% and a diluted system Φ=2.8%. This interaction is repulsive at a distance shorter than the polymer size due to the confinement resistance, and the addition of a polymer-modified at two ends PEO-2m on a microemulsion introduces an attractive interaction between the droplets in the diluted case (Φ=2.8%). On the other hand, the PEO-2m and PEO-m strengthen the repulsive potential between the microemulsions in the concentrated case (Φ=14%).

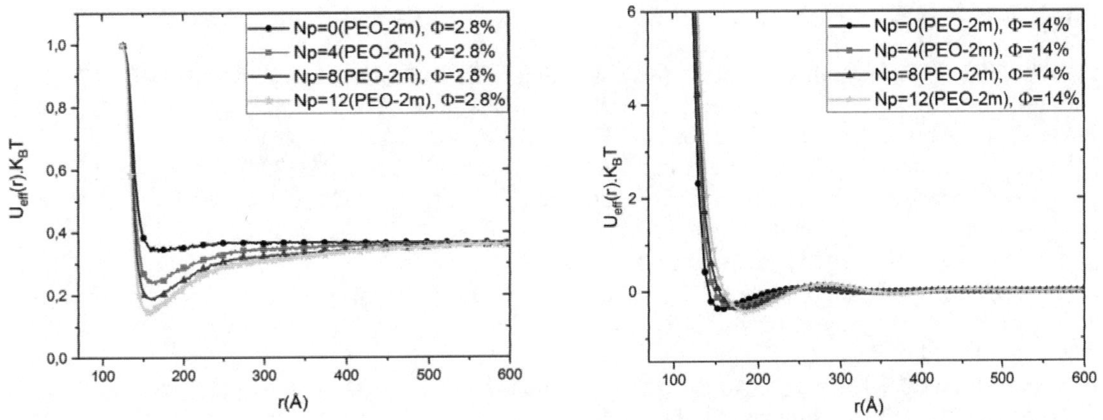

Figure 6. The mean field interaction potential $U_{eff}(r)=-k_BT \log(g(r))$ of microemulsions without PEO-m (Np=0) and with PEO-2m (Np=4, 8 and 12) for Φ=2.8% and Φ=14%.

Conclusion

In this work, we used molecular dynamics simulation to study the structure of oil/water microemulsions with associative polymers (PEO-m and PEO-2m). We were able to compare the effect of the following two polymers: PEO-m and PEO-2m, on the radial distribution function and the mean effective interaction. We notice the addition of PEO-m with different amounts on the bare microemulsion introduces repulsive interactions between the droplets for all volume fractions. On the other hand, the PEO-2m addition to the bare microemulsion introduces an attractive bridging interaction between the droplets in the diluted case. In contrast, it acquaints a repulsive interaction between the droplets in the concentrated case. Then the additions of the polymers POE-2m and POE-m have the same effect on the microemulsion in the concentrated microemulsions systems.

Reference

[1] Mishra A, Panola R and Rana A C 2014 Journal of Scientific and Innovative Research **3** 467-474.
[2] Khatouri M, Ahfir R, Lemaalem M, El khaoui S, Derouiche A and Filali M 2021 RSC Adv **11** 7059.
[3] Lemaalem M, Ahfir R Derouiche A and Filali M 2020 RSC Adv **10** 36155-36163.
[4] Hadrioui N, Lemaalem M, Derouiche A and Ridouane H 2020 RSC Adv **10** 3745-3755.
[5] Lemaalem M, Hadrioui N, El Fassi S, Derouiche A and Ridouane H 2021 RSC Adv **11** 1503-1516.
[6] Srinivasa Rao Y, Deepthi S and Chowdary K 2009 Int. J. Drug Deliv. Technol **1** 39–41.
[7] Nan Y, Li W and Jin Z 2020 Langmuir **36** 5198–5207.
[8] Cohen D, Hitchcock E, and Pohl S 1992 Eur.Patent Appl., EP640 335 A2
[9] Hamaker H C 1937 Physica **4** 1058–1072.
[10]Rowlinson J S 1989 Phys. A **156** 15–34.
[11] Ahfir and Filali M 2020 Adv. Mater. Lett, **11(8)** 20081548.
[12] Belloni L 2000 J. Phys. Condens. Matter **12** 46.
[13] Belloni L 2002 J. Phys. Condens. Matter **14** 9323.
[14] Ahfir R, Azougarh A, Benhamou M and Filali M 2014 International Review of Physics **8** 3.
[15] Filali M, Aznar R, Svenson M, Porte G and Appell J 1999 J.Phys.Chem **103(34)** 7293-7301.
[16] Allen M P 2004. Computational soft matter: from synthetic polymers to proteins **23** 1-28.
[17] Plimpton S J 1995 J. Comput. Phys **117** 1.
[18] Allen M P, Tildesley D 1987 J, Computer Simulation of Liquids, Oxford University Press, Oxford, New York.

IOP Conf. Series: Materials Science and Engineering 1160 (2021) 012010 doi:10.1088/1757-899X/1160/1/012010

First principal calculation and Monte Carlo simulations of the Magnetocaloric effect, Electronic and Magnetic properties in perovskite oxide $Pr_{0.65}Sr_{0.35}MnO_3$

G Kadim[1], R Masrour[1]*, A Jabar[1] and E K Hlil[2]

[1]Laboratory of Solid Physics, Faculty of Science, Sidi Mohammed Ben Abdellah University, Dhar Mahraz, BP 1796, Fez, Morocco
2Univ. Grenoble Alpes, CNRS, Grenoble INP, Institut Néel, 38000 Grenoble, France

E-mail: rachidmasrour@hotmail.com

Abstract. We have used the first principal calculation and Monte Carlo simulations (MCS) to investigate the magnetocaloric effect, electronic and magnetic properties of $Pr_{0.65}Sr_{0.35}MnO_3$ (PSMO) perovskite. The exchange-correlation potential was treated with the generalized gradient approximation (GGA) for the electronic and magnetic properties. The ferromagnetic phase of PSMO is half metallic with 100% spin polarization, which is important in the relation to the colossal magnetoresistance properties of this compound. The magnetic moment is obtained. The thermal variation of magnetization of PSMO has been obtained. The temperature dependence of the magnetic entropy change and the adiabatic temperature are obtained by MCS. The Curie temperature of PSMO has been deduced. The field dependence of relative cooling power PSMO has been obtained.

1. Introduction

The search for energy-efficient technologies for developing new refrigerator appliances has made the magnetocaloric effect a field of current scientific interest [1]. At present, there are several interesting groups of magnetocaloric materials, which are good candidates for active regenerator material used in the room temperature cooling devices. The first group of materials is hydrogenated as well as cobalt doped lanthanum based $La(Fe,Si)_{13}$ intermetallic compounds. The second one is related to the manganese-based compounds, mainly Fe_2P type compounds such as $(Mn,Fe)_2(P,X)$ [X = Ge, Si, As] [2-5]. Magnetic as well as calorimetric measurements have been performed on single crystal samples of $Pr_{0.6}Sr_{0.4}MnO_3$ and $Nd_{0.6}Sr_{0.4}MnO_3$ to develop a complete critical behavior study of the paramagnetic to ferromagnetic transition in both manganites [6]. Previously, the development of a new magnetic refrigeration (MR) technology, based upon the magnetocaloric effect (MCE) [7], has brought an alternative to the conventional gas compression (CGC) technique [8,9]. A thorough understanding of the magnetocaloric properties of existing magnetic refrigerant materials has been an important issue in magnetic refrigeration technology. This paper reviews a new class of magnetocaloric material that is, the ferromagnetic perovskite manganites ($R_{1-x}M_xMnO_3$, where R =La, Nd, Pr and M= Ca, Sr, Ba, etc.) [10-12].

The different research activities can be gathered in the major following axes:

- Study of MCE and research of new materials with high magnetocaloric effect [13-14];
- Study and modeling of thermodynamic cycles [15];
- Design and realization of magnetic refrigeration device with its magnetic source [16–18].

In this spirit, an oxide expected to show promising magnetocaloric properties around room temperature $Pr_{0.65}Sr_{0.35}MnO_3$ was produced in large scale and shaped in order to build a regenerator [19]. on other hand, the perovskite type manganite systems, LnSrMnO (Ln=La, Pr, Nd, Sm, and Gd; $0 \leq x \leq 0.5$) were studied as the electrode materials for solid oxide fuel cells from the viewpoint of applications to the co-firing process of the electrolyte and electrode at a higher temperature [20]. In previous works [21-22], the magnetocaloric effect on $SmFe_{1-x}Mn_xO_3$ and $La_{0.67}Ba_{0.22}Sr_{0.11}MnO_3$ perovskites has been investigated using the Monte Carlo simulations.
In this work, both approaches First principal calculations and Monte Carlo simulations are used to study the $Pr_{0.65}Sr_{0.35}MnO_3$ perovskite. The electronic structure are studied. The thermal of the magnetic entropy change and the adiabatic temperature are determined. The Curie temperature of $Pr_{0.65}Sr_{0.35}MnO_3$ and the field dependence of relative cooling power $Pr_{0.65}Sr_{0.35}MnO_3$ has been obtained.

2. Ab initio calculations

The electronic and magnetic properties of $Pr_{0.65}Sr_{0.35}MnO_3$ were calculated using the density functional theory (DFT) [23-25]. This method is based on the full-potential linearized augmented plane wave method (FPLAPW) [26]. Exchange and correlation effect are treated by the gradient generalized approximation GGA [27]. The structure file of $PrMnO_3$ contains 20 atoms (4 Pr, 4 Mn and 12 O) with lattice parameters a = 5.90(Å), b = 7.72(Å), c = 5.52 (Å) and $\alpha = \beta = \gamma = 90°$ with the space group Pnma [28]. To mimic the realistic concentration of Sr (x = 0.35) in the $PrMnO_3$ Pnma lattice, (1 *1 *5) supercell is formed containing 100 atoms (20 Pr, 20Mn and 60 O atoms in Figure 1).

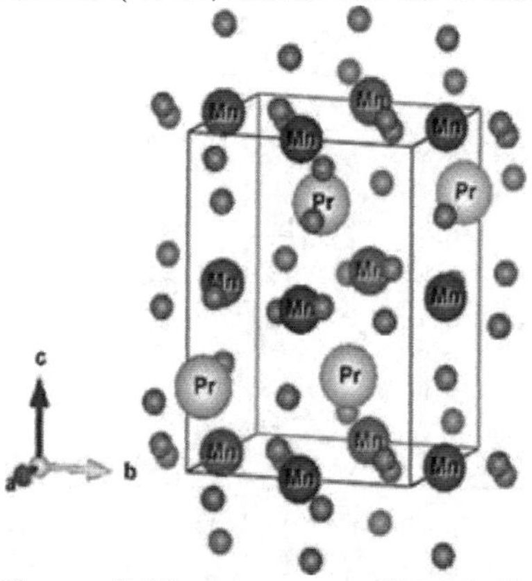

Figure 1. The structure of $PrMnO_3$

A Sr / Pr substitution in the supercell of the total atom of 13 Pr leads to a composition $Pr_{0.65}Sr_{0.35}MnO_3$ (Figure 1), which is even of the composition $Pr_{0.65}Sr_{0.35}MnO_3$ prepared experimentally [19]. The energy of separation between the valence and core states is -9.0 Ry. The valence wave functions inside the muffin-tin spheres are expanded in terms of spherical harmonics up to $l_{max} = 10$. The muffin-tin (MT) radii of Pr, Sr, Mn and O were chosen to be 2.45, 2.23, 1.95 and 1.68 respectively. Both the muffin-tin radius and the number of k-points were varied to ensure total energy convergence. To reduce the calculation time for structural alignment, we based on a distance cutoff 16.281 Bohr.

3. Model and Monte Carlo simulation

The Ising model of $Pr_{0.65}Sr_{0.35}MnO_3$ perovskite is given by:

$$H = -\sum_{<i,j>} J_{ij} S_i S_j - h \sum_i S_i \qquad (1)$$

with J_{ij} is the first, second and third exchange interactions and h is the external magnetic field. The values of J_1=+42.0, J_2=+39.0 and J_3=+36.0 are found from the mean field theory [29]. The spin moment of Mn^{3+} is S=2. The $Pr_{0.65}Sr_{0.35}MnO_3$ perovskite such as given in Figure 1 is assumed to reside in the unit cells and the system consists of the total number of spins N=896. Cyclic boundary conditions on the lattice were applied and the configurations were generated by sequentially traversing the lattice and making single-spin flip attempts. Our data were generated with 10^5 Monte Carlo steps

per spin, discarding the first 10^4 Monte Carlo simulations. Starting from different initial conditions, we performed the average of each parameter and estimate the MCSs, averaging over many initial conditions. The different parameters who calculate are given by the following equations:
The internal energy per site is:

$$E = \frac{1}{N}\langle H \rangle \qquad (2)$$

Magnetizations of material are:

$$M = \frac{1}{N}\left\langle \sum_i \sigma_i \right\rangle \qquad (3)$$

Magnetic entropy is:

$$S(T,h) = \int_0^T \frac{C_m}{T} dT \quad (4)$$

$$\text{With } C_m = \frac{\beta^2}{N}\left(\left\langle E^2 \right\rangle - \langle E \rangle^2\right) \qquad (5)$$

ΔS_m can be calculated indirectly from the experimental magnetization curves by using Maxwell relation [30]:

$$\Delta S(T,h) = \int_0^h \left(\frac{\partial M}{\partial T}\right)_h dh \quad (6)$$

The adiabatic temperature change is given by:

$$\Delta T_{ad} = -T \frac{\Delta S_m}{C_m} \quad (7)$$

The relative cooling power (RCP) described as an area under the dependence of $\Delta S_m(T)$ on temperature, is a compromise between the magnitude of the magnetic entropy change and the width of the peak is given by:

$$RCP = \int_{T_c}^{T_h} \Delta S_m(T) dT \qquad (8)$$

where T_c and T_h are the cold and the hot temperatures corresponding to both ends of the half-maximum value of ΔS_m^{max} ,respectively.

3. Results and discussion

In Figure 2a, we present the total DOS of $Pr_{0.65}Sr_{0.35}MnO_3$ between -7 eV and 5eV as a function of energy. The presence of gap energy close to the Fermi level for spin down shows that this compound has a half metallic character. The summation of the total density of spin up and spin down greater than zero emphasizes that this compound has ferromagnetic behavior. To understand the bonding mechanisms between the atoms, the analysis of partial density of states (PDOS) has been performed Figure 2b. The atoms Mn and O have a significant contribution of the total density in the vicinity of the Fermi level. While Pr has a strong contribution from energy range -0.53eV to 0.85 eV. The contribution from Sr to the total DOS near Fermi level is negligible. Moreover, the Mn atom is dominant by the 3d orbital, and the O atom is dominant by the 2p orbital. Therefore, the strong contribution of Mn-d is due to the hybridization occurring with O(p) states [31]. This suggests that the half-metallicity and the magnetic spin moment are mainly due to the 2p (O) -3d (Mn) coupling. From the Mn-d orbital projected density of states (PDOS) (Figure 2c), one can clearly see that the d_{x2-y2} and d_{xz} orbitals are predominants at E_F, while the d_{xy}, d_{yz} and d_{z2} orbitals are predominants at [-2.36 eV, -1.13 eV]. From O-p we have strong contribution of p_x, p_y and p_z spin up in energy range -6.8 eV to -1.11 eV Figure 2d.

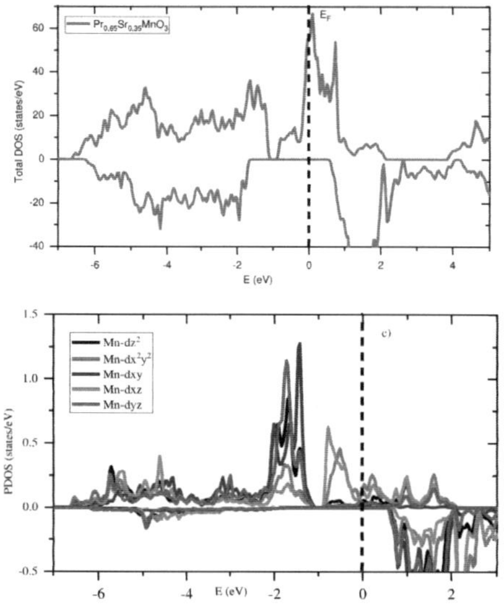

Figure 2. The TDOS (a), PDOS(b), Mn-d orbital projected partial (c) and O-p orbital projected partial (d) of $Pr_{0.65}Sr_{0.35}MnO_3$

We neglected the spin-orbit coupling and using the simple equation: $p(N) = \frac{N\uparrow - N\downarrow}{N\uparrow + N\downarrow}$, where N is the density of states at E_F for spin up ↑or spin down ↓ [32, 33]. It is found that this system present 100% spin polarization. The calculated Mn spin magnetic moment is $3.37\mu_B$ calculated by GGA-PBE (Table 1).

Table 1. the values of magnetic moment of Mn, and spin polarization (SP)

Compound	Magnetic moment (μ_B)	SP (%)
$Pr_{0.65}Sr_{0.35}MnO_3$	3.37617	100%

We have presented in Figure 3 the thermal magnetization obtained by Monte Carlo simulations. From this curve, we deduce that the Curie temperature is equal to T_C=294 K. This value is in good agreement with the experimental one in ref [19] (see Table 2). The magnetization curve shows a classical phase transition of the 2nd order at T_C (ferromagnetic / paramagnetic).

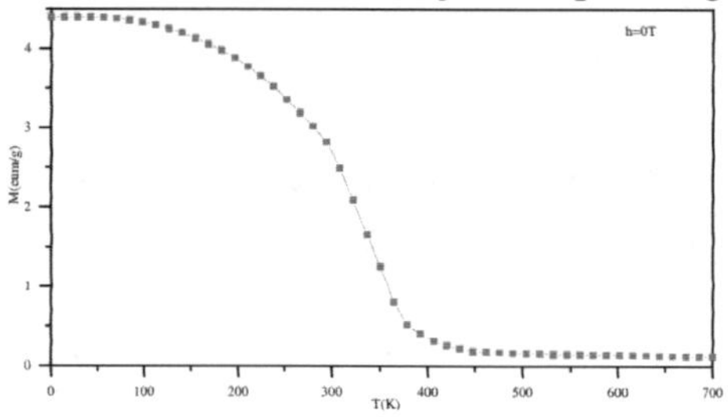

Figure.3. The thermal magnetization for h=0 T

Table 2. The values of T_C, ΔS^{max}, ΔT_{ad} and C_P obtained by Ref.[19] and Monte Carlo simulations.

Perovskite	T_C(K)	$\|\Delta S^{max}\|$ (J/kg.K) at h=1 T	ΔTad(K) at h=1T	C_P(J/(kg.K)) at T_C
$Pr_{0.65}Sr_{0.35}MnO_3$	295 [19]	2.3 [19]	1.1 [19]	580 [19]
	294 Monte Carlo simulations(MCS)	2 MCS	1.13 MCS	581 MCS

Figure 4, shows the variation of thermal magnetic entropy change of $Pr_{0.65}Sr_{0.35}MnO_3$ obtained by Monte Carlo simulations. The value of ΔS^{max} increases with increasing the values of external magnetic field increases while the Curie temperature remains stable. The maximum of magnetic entropy is -ΔS^{max} =1.98 J.K^{-1}.kg^{-1}. This value is in good agreement with the experimental one in ref [19] (see Table 2). The large magnetic entropy change peak, which would be different if the sample is either measured with increasing or decreasing magnetic field. This change in peak is not associated with the intrinsic properties of the sample but to a wrong application of Maxwell relation.

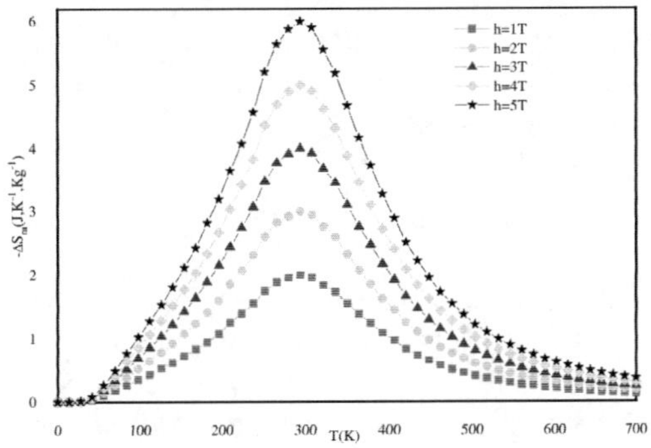

Figure 4. The magnetic entropy change -ΔS_m for a several magnetic fields.

The thermal adiabatic temperature change ΔT_{ad} and heat specific C_P obtained by Monte Carlo simulations are given in Figure 5 for h=1 T. The maximum of adiabatic temperature change (Figure 5a) and heat specific (Figure 5b) are situated at the transition temperature T_C=294 K. This value of T_C is confirmed by the magnetization curve (Figure 3) and magnetic entropy change curve (Figure 4) which exhibits a clear transition around 294 K. The values obtained of ΔT_{ad} and C_P by Monte Carlo are comparable with that given by experimental results of Ref. [19] (see Table 2).

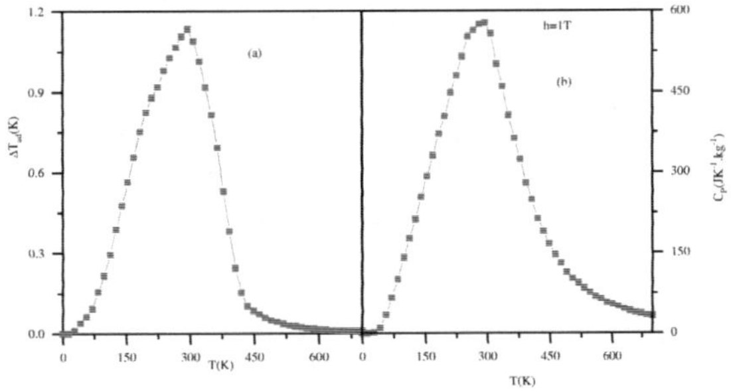

Figure 5. The adiabatic temperature change ΔT_{ad} (a) and heat specific C_P(b) for h=1 T.

We give in Figure 6, the variation of relative cooling power with the magnetic field for $Pr_{0.65}Sr_{0.35}MnO_3$ using Monte Carlo simulations. RCP varies linearly with magnetic field h. The maximum value of RCP is 17557 J/kg is given for h=5T.

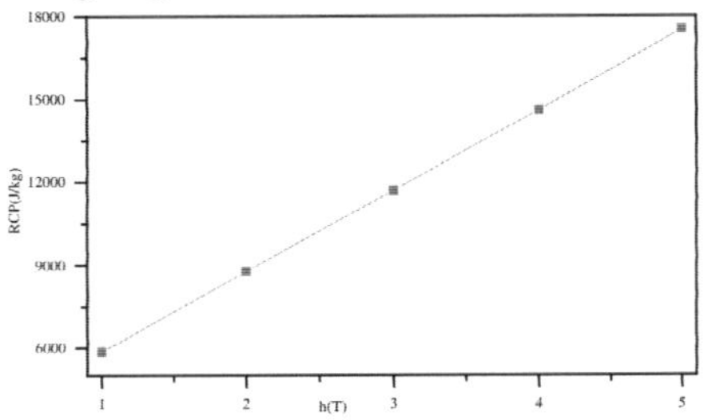

Figure 6. The field dependence of relative cooling power for $Pr_{0.65}Sr_{0.35}MnO_3$.

It is very interesting to note here that usually, this type of giant magnetocaloric effect and large RCP is observed for samples possessing heavier rare-earth elements whose magnetic moments are large [34].

Finally, we have given in Figure 7, the thermal dependence of relative cooling power for $Pr_{0.65}Sr_{0.35}MnO_3$ for a several magnetic fields. The values of RCP increase with increasing the values of magnetic field for a fixed value of temperature. The RCP value increases with increasing the values of temperatures until reached their saturation for each value of magnetic field.

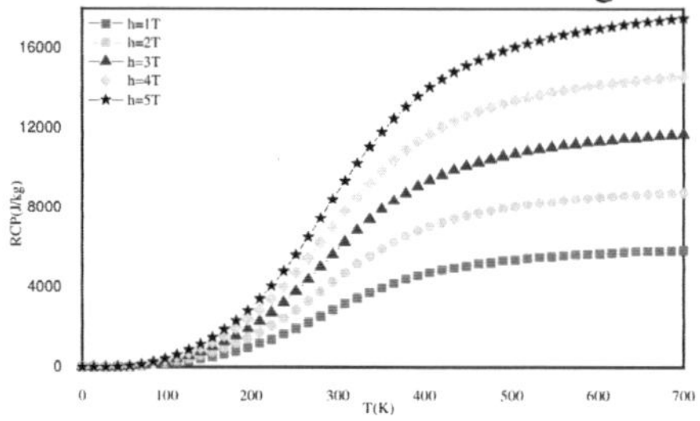

Figure 7. The thermal dependence of relative cooling power for a several magnetic fields.

4. Conclusion

We investigated the magnetocaloric effect, electronic and magnetic properties of $Pr_{0.65}Sr_{0.35}MnO_3$ by using first principle studies and Monte Carlo simulation. The spin polarization for $Pr_{0.65}Sr_{0.35}MnO_3$ show half metallic ground state with the ferromagnetic coupling of Mn spin. The transition temperature has been obtained from the variation of magnetization versus the temperatures for a fixed value of magnetic field. The obtained values of T_C=294 K using Monte Carlo simulations are comparable with that obtained by Ref. [19]. The obtained results of magnetic entropy change and heat specific for Monte Carlo simulations are comparable with that obtained by Ref. [19]. The maximum of magnetic entropy change is situate at the transition temperature and increase with increasing the magnetic field. The transition temperature of $Pr_{0.65}Sr_{0.35}MnO_3$ is obtained and it is comparable with that found by experiment results. The adiabatic temperature and heat specific are obtained. The maximum of curves are situated at the transition temperature. The variation of RCP values versus the magnetic fields and temperatures values are found. The RCP values increase with increasing the magnetic field.

References

[1] Francco V, Blazquez, JS, Ingale B and Conde A 2012 The magnetocaloric effect and magnetic refrigeration near room temperature: materials and models. *Annual Review of Materials Research*, **42**:305-342.
[2] Tegus O, Bruck E, Zhang L, Dagual, Buschow K H J and de Boer F R 2002 *Physica B* **319** 174
[3] Li X W, Tegus O, Zhang L, Dagula, W, and Brück E 2003 *IEEE transactions on magnetics* **39** 3148
[4] Tegus O, Bruck E, Li X W, Zhang L, Dagula W, De Boer F R, and Buschow K H J 2004 *J. Magn.Magn. Mater* **272** 2389
[5] Tegus O, Bruck E, Zhang L, Buschow K H J and De Boer F R 2002 *Physica B* **319** 174
[6] Oleaga A, Salazar A, Hatnean M C and Balakrishnan G 2015 *Phys. Rev.B* **92** 024409
[7] Warburg E 1881. Magnetische untersuchungen. *Annalen der Physik*, **249** 141
[8] Gschneidnerjr K A, Pecharsky V K and Tsokol A O 2005 *Reports on progress in physics*, **68** 1479
[9] Bruck E 2005 *J. Physics D: Appl. Phys* **38**: R381
[10] Phan M H and Yu S C 2007 *J. Magn.Magn. Mater* **308** 325
[11] Kadim G, Masrour R, Jabar A and Hlil E K 2021*Physica A* **573** 125936
[12] Kadim G, Masrour R and Jabar A 2021 *Mater. Today. Commun* **26** 102071
[13] De Oliveira N A and Von Ranke P J 2010 *Phys. Reports* **489** 89
[14] Leitao J V and Bruck E 2014 *Results in Physics*, **4** 31
[15] Nielsen K K, Tusek J, Engelbrecht K, Schopfer S, Kitanovski A, Bahl C R H and Poredos A 2011 *Int. J. Refrigeration* **34** 603
[16] Yu B, Liu M, Egolf P W and Kitanovski A 2010 *Int. J. Refrigeration* **33** 1029
[17] Bahl C R H, Engelbrecht K, Eriksen D, Lozano J A, Bjørk R, Geyti J and Pryds N 2014 *Int. J. Refrigeration* **37** 78
[18] Sari O and Balli M 2014 *Int. J. Refrigeration* **37** 8

[19] Guillou F, Leagait U, Keddous L and Hardy V 2012 *EPJ Web of Conferences*. EDP Sciences, **29** 00021
[20] Sakaki Y, Takeda Y, Kato A, Imanishi N, Yamamoto O, Hattori M and Esaki Y 1999 *Japan Society of Powder and Powder Metallurgy* **46** 293
[21] Elyacoubi A S E, Masrour R and Jabar A 2018 *Solid State Commun* **271** 39
[22] Masrour R, Jabar A and Khlifi H, Jemaa F B, Ellouze M and Hlil E K 2017 *Solid State Communications* **268** 64
[23] Kadim G, Masrour R and Jabar A 2021 *International Journal of Energy Research*. https://doi.org/10.1002/er.6490.
[24] Kadim G, Masrour R and Jabar A 2020 *J. Magn.Magn. Mater* **499** 166263
[25] Kadim G, Masrour R and Jabar A 2021 *Physica Scripta* **96** 045804
[26] Sinh DJ 1994 Planewaves, *Pseudopotentials and the LAPW Method* (US :Springer) chapter 3 pp 23-41
[27] Perdew JP, Burke K and Ernzerhof M 1996. *Phys Review Lett* **77** 3865
[28] Dyakonov V, Bazela W, Duraj R, Dul M, Kravchenko Z, Zubov E and Szymczak H 2013 *Low Temperature Phys* **39** 351
[29] Holland, W E and Brown H A 1972 *Physica Status Solidi (a)*, **10** 249
[30] Gschneidnerjr K A, Pecharsky V K, Tsokol A O 2005 *Reports on Progress in Physics*, **68** 1479
[31] Chakraborty M, Pal P and Sekhar B R 2008 *Solid state commun* **145** 197
[32] Galanakis I and Mavropoulos P 2007 *J.: Condensed Matter* **19** 315213
[33] Naher M I, Parvin F, Islam A K M A and Naqib S H. 2017 *Eur Phys. J. B.* 91: 289-302.
[34] Tishing A M and Spichkin Y I 2003 The Magnetocaloric Effect and Its Applications. Inst. Phys. Publ., (Bristo: CRC Press) p 476

Structural and optical properties of CdO-ZnO thin films synthesized by sol-gel spin coating method

A Nfissi[1], M Belhajji[1], L Mrharrab[1,2], Y Ababou[1,3] and S Sayouri[1]

[1] LPAIS, Faculty of Sciences-DM, USMBA, B.P. 1796, Fez-Atlas, Morocco
[2] ERMAM, FP, Ouarzazate, Morocco

[3] Corresponding author: ababouyahya@hotmail.com

Abstract. CdO-ZnO thin films with volume ratio of (Cd: Zn) of (1: 0), (0.75: 0.25), (0.5: 0.5) and (0: 1) were prepared using the sol-gel spin coating method. The as-coated films were annealed at 550 °C for 1h and their structural and optical properties investigated. From structural investigations, it was observed that ZnO phase is present in the composites. The crystallite size values decreased with increasing the Zn content, and the dislocation density increased from 0.77 to 0.988 (10^{15} lines/m^2). The optical properties were examined by a ultraviolet-Visible spectroscopy, which indicated that the optical transmission increases in the visible region, and the band gap is found to be blue shift from 2.26 to 3.24 eV with the increase of ZnO content.

1. Introduction

Cadmium oxide (CdO) is an n-type semiconductor with a cubic NaCl-type crystal structure, possesses a direct band gap of 2.2-2.7 eV [1-3]. In addition, CdO films shows high transparency especially in the NIR spectral region, and have been used for many applications such as transparent conducting oxide (TCO), optical communications, photovoltaic solar cells, photo transistors, gas sensors devices and thin-film resistors [4-8]. CdO thin films can be synthesized by different chemical and physical deposition methods such as sol-gel [9], spray pyrolysis [10], magnetron sputtering [11], chemical bath deposition [12], pulsed laser deposition [13] and electrodeposition [14].

Zinc oxide (ZnO) is an n-type semiconductor with a wide band gap in the near ultraviolet of 3.3 eV [15]. In addition, ZnO shows interesting properties such as good transparency, high thermal conductivity and high electron mobility. Besides that, ZnO have been used in light-emitting diodes (LED), solar cells, photodetectors and piezoelectric devices [16-18].

Optical properties of nanostructured Zn and Cd oxides mixture films can be controlled by the oxides content ratio [19]. Caglar et al. [20] prepared CdZnO films with various compositional ratios using the sol-gel method and have found that the optical transmission increased with the increase of Zn concentration.

In the present work, we report elaboration and characterization of CdO-ZnO thin films. The main interest is to study the effect of ZnO content on the structural and optical properties of CdO thin films.

2. Experiment

The following precursors were needed to elaborate the CdO-ZnO thin films: cadmium acetate dihydrate ($Cd(CH_3COO)_2,2H_2O$) and zinc acetate dihydrate ($Zn(CH_3COO)_2,2H_2O$). During the preparation process monoethanolamine (C_2H_7NO, MEA) and 2-methoxyethanol ($C_3H_8O_2$) were used as a stabilizer and solvent.

CdO and ZnO solutions of 0.5M were synthesized separately: at first, cadmium acetate dihydrate was dissolved in 2-methoxyethanol, then, MEA was added to the solution and the resulting mixture was stirred at 60°C for 3h. ZnO solution was synthesized by the same method (stirred at 70°C for 3h). The resulting CdO and ZnO solutions were mixed in various volume ratios of (Cd: Zn) of (1: 0), (0.75: 0.25), (0.5: 0.5) and (0: 1). These CdO-ZnO Composites were denoted S_0, S_{25}, S_{50} and S_{100} respectively, and were stirred at 60°C for 30 min. The final solutions were deposited on the glass substrates, and then rotated at 3000 rpm for 20 s using a spin coater, the films were pre-heated at 300 °C for 10 min in order to evaporate the solvent, these operations were repeated 10 times in order to increase the thickness of the thin films. Finally, the prepared CdO-ZnO thin films were annealed at 550 °C for 1h in a tube furnace in air ambiance.

The structural and optical characterizations of the samples were performed using an X-ray diffraction (XPERT-PRO diffractometer) with Cu-Kα radiation (λ=1.5405980 Å), ultraviolet-Visible (UV-Vis) spectroscopy was used at room temperature in the wavelength range from 300 to 700 nm, respectively.

3. Results and discussion

3.1. Structural analysis

The room temperature X-ray diffraction (XRD) patterns of CdO-ZnO thin films with different Cd: Zn ratios are plotted in figure 1.For the CdO sample (S_0) exhibits (111), (200), (220), (311) and (222) characteristic peaks of cubic NaCl-type structure, and the calculated lattice parameter is 4.69 Å, this is in good agreement with the result reported [21]. The characteristic peaks (100), (002) and (101) of the ZnO sample (S_{100}) were indexed to hexagonal wurtzite structure, and the calculated lattice parameters are a=3.20 Å, c=5.22 Å. For S_{25} and S_{50} samples, peaks of CdO and ZnO are present.

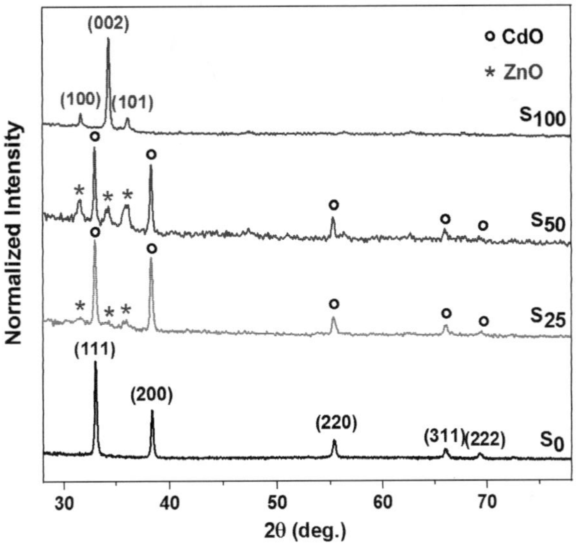

Figure 1. XRD patterns of CdO–ZnO thin films with different Cd:Zn ratios: S_0, S_{25}, S_{50} and S_{100}.

The crystallite size (D) of CdO-ZnO thin films was estimated from the peak (111), using Scherrer's equation:

$$D = \frac{0.9\,\lambda}{\beta\cos(\theta)} \tag{1}$$

where β is the full-width-at-half-maximum (FWHM) of a characteristic diffraction peak, θ is the diffraction angle and λ=1.5406 Å.

The crystallite size values decreased from 36.01 nm to 31.81 nm, with increasing the Zn content from 0 to 50%. This can be ascribed to the smaller ionic radius of Zn^{2+} (0.74 Å) than that of Cd^{2+} (0.95 Å). The dislocation density (δ) of CdO-ZnO thin films was calculated from the crystallite size using the equation:

$$\delta = \frac{1}{D^2} \tag{2}$$

The small dislocation density (0,7711.10^{15} lines/m^2) for pure CdO means that the crystallization of this thin films is better. It is also seen that the δ value of CdO-ZnO thin films is increased with increasing the Zn content from 0 to 50%. This means that ZnO increases the defects in the CdO thin films. This behavior was also observed in Al doped CdO films [22].

The full-width-at-half-maximum (FWHM), the crystallite size (D) and the dislocation density (δ) of CdO-ZnO thin films were shown in Table 1.

Table 1. Structural parameters of CdO–ZnO thin films.

Sample	FWHM of (111) peak (°)	D (nm)	δ (10^{15} lines/m^2)
S_0	0.2299	36.01	0.7711
S_{25}	0.2457	33.70	0.8052
S_{50}	0.2603	31.81	0.9882

3.2. UV-Vis analysis

The room temperature UV-Vis transmittance spectra of CdO-ZnO thin films with different Cd: Zn ratios are plotted in figure 2. It is evident that the transmittance increases with Zn content to reach its maximum value at Cd 0: Zn 1 (i.e. S_{100} sample), then pure ZnO shows high transparency in the visible spectral region in contrast to the pure CdO. This can be explained by the higher value of the surface roughness of the pure CdO thin film due to the increased optical scattering of light incident on the film mince surface [23].

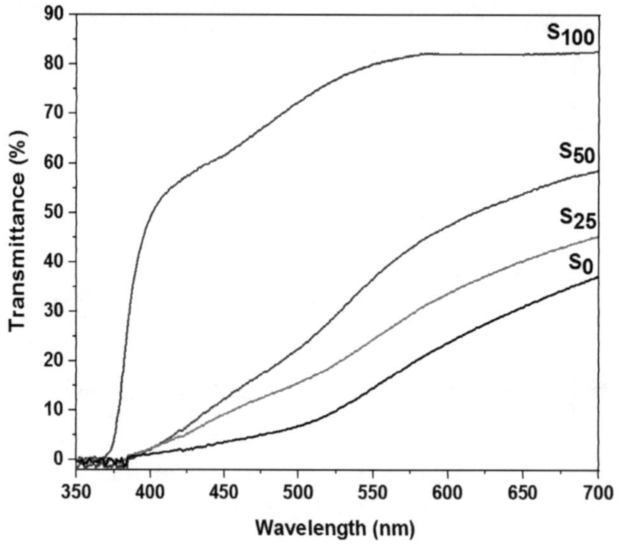

Figure 2. UV-Vis transmittance curves of CdO–ZnO thin films.

The absorbance (A) spectra of CdO-ZnO thin films was calculated from the measured transmittance (T) spectra using the relation:

$$A = -\log_{10}(\frac{T}{100}) \quad (3)$$

The wavelength of the gap energy was extracted by extrapolating the linear part of the curve absorbance, and an illustration of such determination for all the sample is shown on figure 3.

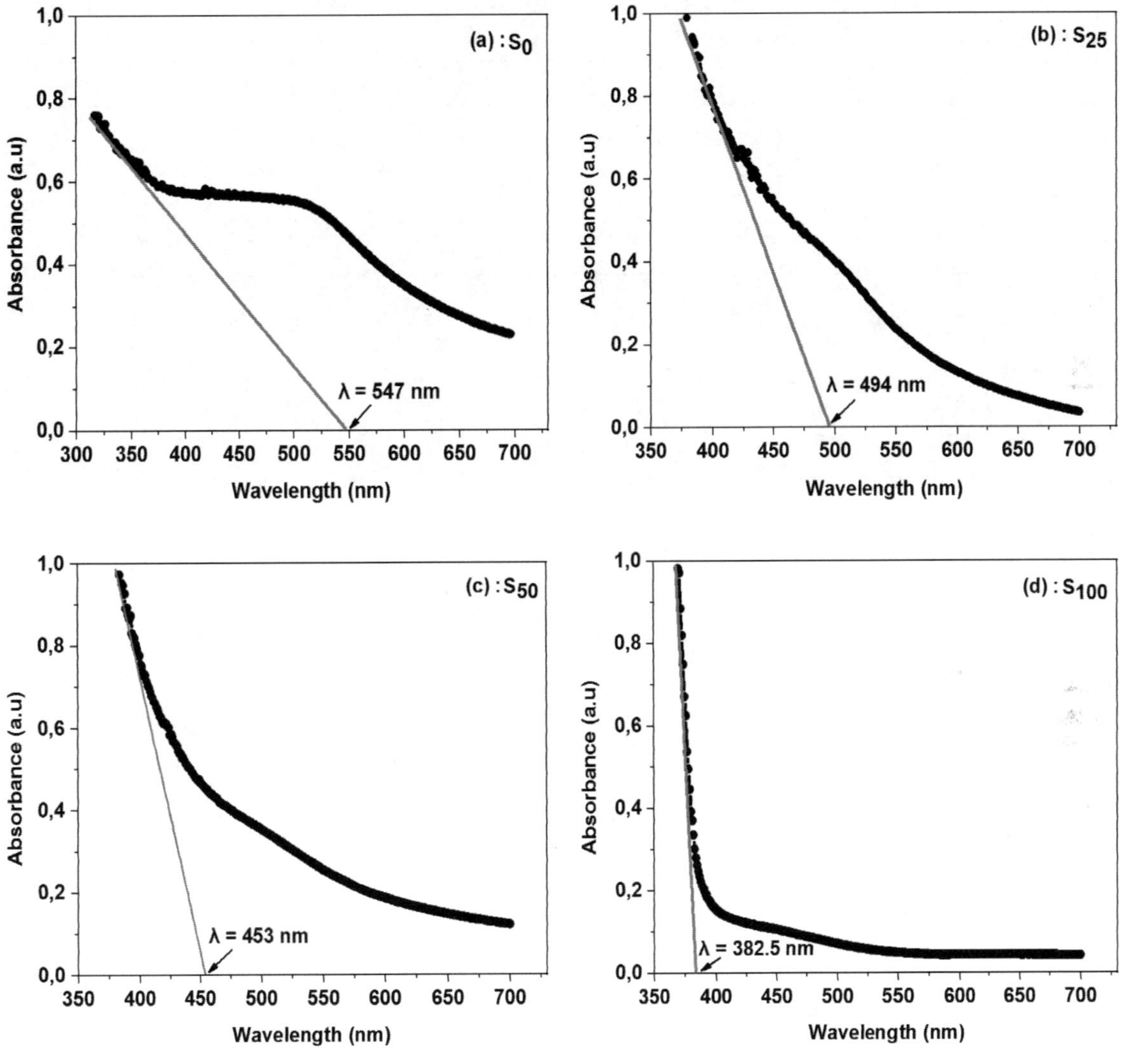

Figure 3. Determination of the wavelength of the gap energy from the absorbance, for CdO–ZnO thin films with different Cd:Zn ratios: (a) S_0, (b) S_{25}, (c) S_{50} and (d) S_{100}.

The optical band gaps of all the sample were calculated from the wavelength using the equation:

$$E_g = \frac{1240}{\lambda} \quad (4)$$

where λ is the wavelength determined from the absorbance spectra.

Table 2 gathers E_g values of all the compositions studied which show a increase with an increase in Zn content. The recorded values of this parameter for S_0 sample belong to the interval of values known for the pure CdO which lay between 2.2 and 2.7. For S_{100} sample a band gap energy of 3.24 eV was obtained, which is in good agreement with reported values of pure ZnO par Singh et al. [24]. A. A. Ziabari et al. [25] prepared CdO-ZnO composite films with volume ratio of (Cd:Zn) of (0:1), (1/4:3/4), (1/2:1/2), (3/4:1/4) and (1: 0) via the sol-gel dip coating method, and have obtained the following E_g values 3.28 eV, 3.03 eV, 2.50 eV, 2.44 eV and 2.23 eV, respectively.

Table 2. Band gap energy of CdO–ZnO thin films.

Sample	Eg (eV)
S_0	2.26
S_{25}	2.51
S_{50}	2.74
S_{100}	3.24

It is observed from figure 4 that E_g increases linearly as a function of Zn content. Hence, the band gap energy exhibits a blue shift with the rise of Zn content from 0 to 100%, which shows the strong influence of the ZnO on the band gap of the CdO thin films. Moreover, the parameter E_g decrease with increasing crystallite size. This behavior was also observed in CdO-ZnO composite films [25].

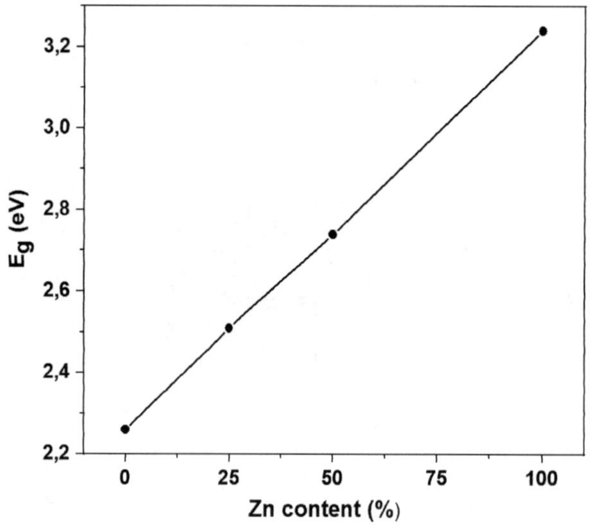

Figure 4. Band gap energy of CdO–ZnO thin films as a function of Zn content.

4. Conclusion

CdO-ZnO thin films were successfully prepared on glass substrates using the sol gel method at a temperature of calcination 550 °C for 1h. Structural analysis showed that pure CdO and ZnO crystallize in the cubic NaCl-type and hexagonal wurtzite structures, respectively. The crystallite size values decreased with increasing the Zn content from 0 to 50%, and the small dislocation density for pure CdO means that the crystallization of this thin film is good. The transmittance increases with increasing the Zn content from 0 to 100% in the visible spectral region. The obtained values of the band gap energy from absorbance spectra shifted to the higher energy as a result of the increasing of Zn content in the thin film.

Nomenclature

a, c Lattice parameters
D Crystallite size
δ Dislocation density
T Transmittance
A Absorbance
λ Wavelength
E_g Band gap energy
FWHM Full-width-at-half-maximum
NIR Near-infrared
UV-Vis Ultraviolet-Visible
TCO Transparent conducting oxide
LED Light-emitting diodes
MEA Monoethanolamine

References

[1] Coutts T J, Young D L, Li X, Mulligan W P and Wu X 2000 *J. Vac. Sci. Technol. A* **18** 2646-60
[2] Ortega M, Santana G and Morales-Acevedo A 2000 *Solid State Electron.* **44(10)** 1765-69
[3] Carballeda-Galicia D M, Castanedo-Perez R, Jimenez-Sandoval O, Jimenez Sandoval S, Torres-Delgado G and Zuniga-Romero C I 2000 *Thin Solid Films* **371** 105-8
[4] Liu X, Xu Z and Shen Y 1997 *in Proc. Int. Conf. on Solid-State Sensors and Actuators* (Lisbon) p 585
[5] Lewis B J and Paine D C 2000 *Mater. Res. Soc. Bull.* **25** 22-7
[6] Yan M, Lane M, Kannewurf C R and Chang R P H 2001 *Appl. Phys. Lett.* **78** 2342-44
[7] Gomez Daza O, Arias-Carbajal Readigos A, Campos J, Nair M T S and Nair P K 2001 *Modern Phys. Lett. B* **15** 609-12
[8] Zhao Z, Morel D L and Ferekides C S 2002 *Thin Solid Films* **413** 203-11
[9] Ghosh P K, Das S, Kundoo S and Chattopadhyay K K 2005 *J. Sol-Gel Sci. Technol* **34(2)** 173–9
[10] Reddy K T R, Shanthini G M, Johnston D and Miles R W 2003 *Thin Solid Films* **427** 397–400
[11] Henríquez R, Grez P, Muñoz E, Gómez H, Badán J A, Marotti R E and Dalchiele E A 2010 *Thin Solid Films* **518** 1774–78
[12] De León-Gutiérrez L R, Cayente-Romero J J, Peza-Tapia J M, Barrera-Calva E, Martínez-Flores J C and Ortega-López M 2006 *Mater. Lett.* **60** 3866–70
[13] Gupta R K, Ghosh K, Patel R, Mishra S R and Kahol P K 2009 *Curr. Appl. Phys.* **9** 673–7
[14] Liu Z Q, Guo R, Li G R, Bu Q, Zhao W X and Tonga Y X 2012 *Electrochim. Acta* **59** 449-54
[15] Guo H, Zhou J and Lin Z 2008 *Electrochem. Commun.* **10** 146-50
[16] Saravanan R, Shankear H, Prakash T, Narayanan V and Stephen A 2011 *Mater. Chem. Phys.* **125** 277–80
[17] Mosquera E, Bernal J, Zarate R A, Mendoza F, Katiyar R S and Morell G 2013 *Mater. Lett.* **93** 326–29
[18] Samanta P K and Chaudhuri P R 2011 *Sci. Adv. Mater.* **3** 107–12
[19] Yousef A, Barakat N A M, Amna T, Unnithan A R, Al-Deyab S S and Kim H Y 2012 *J. Lumin.* **132** 1668-77
[20] Caglar Y, caglar M, Ilican S and Ates A 2009 *J. Phys. D: Appl. Phys.* **42** 065421
[21] Zhao Z, Morel D L and Ferekides C S 2002 *Thin Solid Films* **413** 203–11
[22] Aydemir S, Köse S, Kilickaya M S and Özkan V 2014 *Superlattices Microstruct.* **71** 72-81
[23] Paine D C, Whitson T, Janiac D, Beresford R and Yang C O 1999 *J. Appl. Phys.* **85** 8445-50

[24] Singh G, Srivastava A K, Senthil K, Tawale J S, Deepa M and Yong K 2011 *J. Nanoeng. Nanomanuf.* **1** 77–83
[25] Ziabari A A and Ghodsi F E 2011 *J. Alloys Compd.* **509** 8748–55

AUTHOR INDEX

Ababou, Yahya 23, 32
Ababou, Y 84
Abdi, F 1
Ahfir, R 10, 69
Ahjyaje, F Z 1
Amarass, M 1
Amila, I 17
Bakkari, K El— 54
Belhajji, M 84
Bih, L 63
Bouftila, Nour El Hoda 23
Chouiekh, Abdelhak 32
Elkhaoui, S 10, 69
Farid, Abdi 41
Fedoul, A 17
Filali, M 10, 69
Haddad, M 63
Hlil, E K 77
Idrissi, M Janati 17
Iqraoun, E 54
Jabar, A 77
Jerroudi, M 63
Kadim, G 77
Khatouri, M 10, 69
Lamcharfi, T 1
Lemaalem, M 10, 69
Malika, Ahabboud 1
Masrour, R 77
Mrharrab, L 84
Mustapha, Amarass 41
Naji, Mohamed 23, 32
Najwa, Gouitaa 41
Nfissi, A 84
Nor-Said, Echatoui 41
Rjeb, Abdelilah 23, 32
Saadoune, I 63
Sali, A 54
Sayouri, Salaheddine 23, 32
Sayouri, S 17, 84
Taj-Dine, Lamcharfi 41
Talha, L 10, 69